Manufacturing
Mathematical Models, Problems, and Solutions

T0141229

Manufacturing
Mathematical Models, Problems, and Solutions

Zainul Huda
Professor in Manufacturing Technology
Department of Mechanical Engineering
King Abdulaziz University

CRC Press
Taylor & Francis Group
Boca Raton London New York

CRC Press is an imprint of the
Taylor & Francis Group, an **informa** business

CRC Press
Taylor & Francis Group
6000 Broken Sound Parkway NW, Suite 300
Boca Raton, FL 33487-2742

First issued in paperback 2021

© 2018 by Taylor & Francis Group, LLC
CRC Press is an imprint of Taylor & Francis Group, an Informa business

No claim to original U.S. Government works

ISBN-13: 978-0-367-78127-9 (pbk)
ISBN-13: 978-1-138-50136-2 (hbk)

Visit the Taylor & Francis Web site at
http://www.taylorandfrancis.com

and the CRC Press Web site at
http://www.crcpress.com

Contents

Section I   Basic Concepts and Manufacturing Processes

Section II   Advanced Manufacturing

Section III   Quality and Economics of Manufacturing

List of Figures

List of Tables

Preface

This is not just a problems and solutions book; rather this volume is designed to cover three aspects of manufacturing technology: (a) fundamental concepts, (b) engineering analysis/mathematical modeling of manufacturing operations, and (c) more than 250 problems and their solutions. These attractive features render this book suitable for recommendation as a textbook for undergraduate- as well as master's-level programs in mechanical/materials/ manufacturing/industrial engineering. This unique book is equally useful to both engineering-degree students and design/production engineers practicing in the manufacturing industry.

The book is divided into three parts. Part I introduces readers to manufacturing and basic manufacturing processes (metal casting, plastic molding, metal forming, heat treatment, surface engineering, weld design and joining, powder metallurgy, ceramic processing, and composite processing) and their mathematical modeling with diagrammatic illustrations followed by worked examples. Part II covers nontraditional machining and computer-aided manufacturing, including their mathematical modeling and the related solved problems. Finally, quality assurance and economic aspects of manufacturing are dealt with in Part III.

The engineering analysis/mathematical modeling of a manufacturing operation is so important that it is almost impossible for a manufacturing/material process engineer to produce a defect-free part without solving/programming the involved numerical problems. For example, in order to manufacture a defect-free sand-cast product, an engineer must properly design pattern, sprue, and riser by using the relevant mathematical models and by solving the casting design problems. Similarly, the computations for material removal rate (MRR) and cutting time (T_c) are also industrially important for assessing manufacturing efficiency and the completion of machining job well in time.

Computer aided manufacturing (CAM) and non-traditional machining processes are discussed in Part Two. Here, computer aided design (CAD), computer numerical controlled (CNC) machining, ultrasonic machining (USM), abrasive water jet machining ($AWJM$), AJM, electric discharge machining (EDM), electron beam machining (EBM), and electrochemical machining (ECM) have been mathematically modeled; and solution to the problems (including calculations for MRR) are presented. Finally, chapters on problems and solutions on economic and quality engineering aspects are included so as to develop the skills in readers to solve problems independently and to manufacture quality-assured and cost-effective products for complying with market requirements.

There are a total of 19 chapters. Each chapter first introduces readers to the technological importance of the chapter topic, provides definitions of terms and their explanations, and then presents mathematical modeling/ engineering analysis. The meanings of the terms along with their SI units in each mathematical model are clearly stated. There are over 320 mathematical models/equations. An attempt is also made to provide diagrammatic illustrations for each manufacturing process and its mathematical model. The applications of mathematical models are illustrated with the aid of *Examples* (solved quantitative problems) at the end of each chapter. There are a total of 250 solved problems (*Examples*); each *Example* clearly mentions the specific formula applied to solve the problem followed by the steps of solution. Additionally, *Exercise Problems* are included for every chapter. The answers to selected problems are also provided at the end of the book.

Zainul Huda
Department of Mechanical Engineering
King Abdulaziz University

Acknowledgments

By the grace of God, I have been able to complete the write-up of this book's manuscript and get it published. I am deeply thankful to my wife for her continued patience during many late nights while writing this book. I would like to thank Abdel Salaam, PhD, Mechanical Engineering (University of Washington), Associate Professor, Department of Mechanical Engineering King Abdulaziz University, Jeddah, Saudi Arabia, for his valuable suggestions on *weld design*. I am also grateful to Osama Khashaba, Professor, Mechanical Engineering Department, King Abdulaziz University, for the valuable discussion with him on composite materials processing.

I acknowledge Sarfraz Hussain, Vice Chancellor, DHA Suffa University, Pakistan, for granting permission to publish the photograph of the CNC machine in the *CIM* (Computer Integrated Manufacturing) Centre of his institution. I am also indebted to Muhammad Nauman Qureshi, Head of the Department of Mechanical Engineering, DHA Suffa University, for providing the photograph and technical details of the CNC machine. I am grateful to Tuan Zaharinie, Senior Lecturer, Mechanical Engineering Department, University of Malaya, Kuala Lumpur, Malaysia, for contributing a problem and its solution on metal casting. Last, I would like to express my appreciation to Saleh Al-Ghamdi, senior year student, Department of Mechanical Engineering, King Abdulaziz University, for developing the drawing using AutoCAD for casting design.

Author

Zainul Huda is the author/co-author/editor of five books, in addition to this text, and over 120 research/academic articles, including the book titled *Materials Processing for Engineering Manufacture* (2017, Trans Tech Publications, Switzerland). Huda is a professor in manufacturing technology at the Department of Mechanical Engineering, King Abdulaziz University, Saudi Arabia. His teaching interests include manufacturing technology, material removal processes, materials science, aerospace materials, metallurgy, and failure analysis. He has been working as a professor at reputed universities (e.g., University of Malaya, Malaysia; King Saud University, Saudi Arabia, etc.) since February 2007. He has over 35 years of academic experience in materials/metallurgical/mechanical engineering at various universities in Malaysia, Pakistan, and Saudi Arabia.

Huda earned a PhD in materials technology from Brunel University, London, United Kingdom, in 1991. He is a postgraduate in manufacturing engineering. He obtained a B.Eng. in metallurgical engineering from the University of Karachi, Pakistan, in 1976. He is the author/co-author/editor of more than 125 publications, including five books and 32 peer-reviewed international SCOPUS-indexed journal articles (26 ISI-indexed papers) in the fields of materials/manufacturing/mechanical engineering, published by renown publishers from the United States, Canada, the United Kingdom, Germany, France, Switzerland, Pakistan, Saudi Arabia, Malaysia, South Korea, and Singapore. He has been cited over 455 times in Google Scholar. His author *h*-index is 11 (*i*10-index: 13). As principal investigator (*PI*), he has attracted seven external and internal research grants; these research funds' amount to a total close to $100,000. He is the recipient of the Vice Chancellor's Appreciation Certificate awarded by the vice chancellor, University of Malaya, for an invention/patent.

Huda has successfully completed more than 20 industrial consultancy/research and development projects in the areas of failure analysis and manufacturing in Malaysia and Pakistan. He is the developer of Toyota Corolla cars' axle-hub's heat treatment procedure first ever implemented in Pakistan (through Indus Motor Co./Transmission Engineering Industries Ltd, Karachi) in 1997. In addition to an industrial consultancy, Huda has worked as C. plant manager, development engineer, metallurgist, and Gr. engineer (metallurgy) in various manufacturing companies, including Pakistan Steel Mills Ltd.

Huda has delivered guest lectures/presentations in the areas of materials and manufacturing in the United Kingdom, South Africa, Saudi Arabia, Pakistan, and Malaysia. He holds memberships in prestigious professional societies, including the Pakistan Engineering Council; the Institution of Mechanical Engineers, London, United Kingdom; and the Canadian Institute of Mining, Metallurgy, and Petroleum, Canada. He has supervised/co-supervised a number of PhD and master's research theses. His biography has been published in Marque's *Who's Who in the World, 2008–2016*. He is a member of the International Biographic Centre, UK's *Top 100 Engineers – 2015*. Huda is the recipient of the UK-Singapore Partners in Science Collaboration Award for research collaboration with British universities, which was awarded to him by the British High Commission, Singapore in 2006. He is a Google Scholar's World's Top Ten Research Scholar in the field of materials and manufacturing (see https://scholar.google.com/citations?view_op=search_authors&hl=en&mauthors=label:materials_and_manufacturing&before_author=8Vfd_zYAAAAJ&astart=0).

Section I

Basic Concepts and Manufacturing Processes

.

1

Introduction

1.1 Manufacturing Process Model

Manufacturing involves the transformation of raw materials into finished or semi-finished products; this activity can be presented as a model of the manufacturing process (see Figure 1.1). It is evident from the manufacturing model that machine tools, equipment, and plant facilities are housed in a suitable building, which must have an adequate supply of utilities (water, electricity, fuel, etc.) to run the machine tools or processing equipment. Additionally, there must be, of course, a workforce to perform the manufacturing operations. An efficient plant layout is an important requirement of modern manufacturing.

A manufacturing process usually involves a sequence of manufacturing operations to convert raw materials into finished or semi-finished products (Youssef et al., 2011). A typical example is the powder metallurgy (P/M) process. The P/M process starts with the crushing of lumps (large pieces) of the material into smaller pieces by crushing in a crusher followed by grinding in ball mills to obtain powder (particles size <1 mm) (James, 2015). The powder (fine particles) is then screened and blended. The blended powder is then compacted in a press to obtain green compacts which are sintered (heated) to obtain the product.

1.2 Manufacturing Processes and Technologies

Manufacturing processes and technologies may be classified into three groups: (1) basic manufacturing processes, (2) nontraditional manufacturing/machining processes, and (3) advanced manufacturing technologies. Each of these manufacturing categories has various types; which are explained in the following subsections.

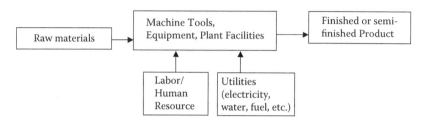

FIGURE 1.1
A model of a manufacturing process.

1.2.1 Basic Manufacturing Processes

Basic manufacturing processes refer to the group of conventional processes that are generally used for manufacturing tangible products. There are eight basic or conventional manufacturing processes: (1) metal casting, (2) metal forming, (3) machining, (4) welding/joining, (5) heat treatment, (6) surface finishing/engineering, (7) powder metallurgy, and (8) plastic/ceramic molding.

Metal Casting: Metal casting involves mold design and preparation, melting of metal in a furnace, and admitting molten metal into the mold cavity (where the metal solidifies at a controlled rate and takes the shape of the mold cavity) followed by the removal of the solidified metal (casting) from the mold. In most cases the casting is cleaned, heat treated, finished, and quality assured before shipping to market. There are a variety of different types of casting processes. The most common example of metal casting is the sand-casting process. Metal casting processes and their mathematical modeling are discussed in detail in Chapters 2 and 3.

Metal Forming: The plastic deformation of a metal into useful size and shape is called metal forming. Depending on the forming temperature, metal forming may be either hot working or cold working. There is a wide variety of different types of metal forming processes (Creese, 1999). A typical example of a metal forming process is *rolling*. In *metal rolling*, the metal stock is passed through one or more pairs of rolls to reduce the thickness or to obtain the desired cross section (see Figure 1.2). Metal forming processes and their mathematical modeling are discussed in Chapters 5–7.

Machining: Machining involves the removal of material from a workpiece by using a cutting tool. Machining can be considered as a system consisting of the workpiece, the cutting tool, and the equipment (machine tool). In conventional or traditional machining, material removal is accompanied by the formation of chips, which is accomplished by use of either a single-point or a multiple-point cutting tool. Nontraditional or advanced machining are chip-less material removal processes that involve the use of energy (rather than a cutting tool) for material cutting. Further description and mathematical modeling of machining processes are given in Chapters 8–10 and in Chapter 16.

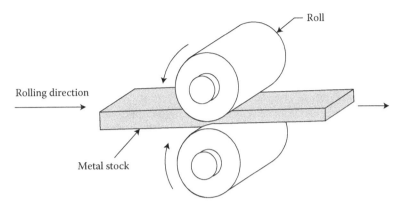

FIGURE 1.2
Rolling—a bulk-deformation metal forming process.

Welding/Joining: Welding ranks high among industrial manufacturing processes. Joining two or more pieces of metal to make them act as a single piece by *coalescence* is called *welding*. In general, metals are permanently joined together by one of the following three technologies: (1) welding, (2) brazing, and (3) soldering. Among the three joining technologies, welding is the most important, the most economical, and the most efficient way to permanently join metals. The welding of two workpieces usually requires them to be heated to their melting temperatures (Huda, 2017). Brazing and soldering involves the joining of two workpieces at a temperature below their melting points. Many different energy sources can be used for welding; these include an electric arc, a gas flame, a laser, an electron beam, friction, ultrasound, and the like. Depending on the type of energy source and techniques, a wide variety of (more than 50) welding processes have been developed. Weld design and welding processes are explained in Chapter 13.

Heat Treatment: This involves the controlled heating, holding, and cooling of metals for altering their physical and mechanical properties. Depending on the mechanical properties to be achieved, heat treatment processes for steels can be classified into (1) full annealing, (2) normalizing, (3) hardening and tempering, (4) recrystallization or process annealing, and (5) spheroidizing annealing. These heat treatment processes of steel involve the application of a iron-iron carbide phase equilibrium diagram and a time-temperature-transformation diagram. In *hardening* heat treatment, steel is slowly heated at a temperature about 50°C above the *austenitic temperature* line A_3 or line A_{CM} in the case of hypo-eutectoid steels and 50°C above A_1 line into the austenite-cementite region in the case of hypereutectoid steels. The steel is held at this temperature for a sufficiently long time to fully transform phases into either austenite or austenite + cementite as the case may be. Finally, the hot steel is rapidly cooled (quenched) either in water (to form martensite) or in oil (to form bainite) so as to ensure high hardness. In many automotive components, hardening is followed by tempering. This author has ensured the required high hardness in a forged hardened-tempered AISI-1050 steel for application in the axle-hubs of a motorcar by achieving a tempered martensitic microstructure (Huda, 2012).

The scope of this book does not allow us an in-depth discussion on heat treatment; the latter is given elsewhere (Dossett and Boyer, 2006; Huda, 2017).

Surface Engineering: Surface engineering, also called *surface finishing/coating,* aims at achieving excellent surface finish and/or characteristics by use of either finish grinding operations or by plating and surface coatings to finish part surfaces. Applied as thin films, these surface coatings provide protection, durability, and/or decoration to part surfaces. The most common plating and surface coating technologies used include (1) painting, (2) electroplating, (3) electroless plating, (4) conversion coating, (5) hot dipping, and (6) porcelain enameling. Surface engineering is discussed in detail in Chapter 11.

Powder Metallurgy (P/M): This comprises a sequence of manufacturing operations involving metal powder production, and their mixing and blending, compaction, and sintering. In P/M, a feedstock in powder form is processed to manufacture components of various types, which include bearings, watch gears, automotive components, aerospace components, and other engineering components. Powder metallurgy is discussed in detail in Chapter 12.

Plastic/Ceramic Molding: It is the manufacturing process of shaping plastic using a mold or rigid frame. The plastic/ceramic molding process enables us to produce plastic/ceramic objects of all shapes and sizes with great design flexibility for both simple and complex designs. A variety of plastic/ceramic molding processes and technologies have been developed to process both thermoplastic and thermosetting plastic materials for various household and engineering applications. The most common technique to produce plastic/ceramic parts is the injection molding process. A detailed explanation of plastic molding processes is given in Chapter 4.

1.2.2 Nontraditional Machining Processes

Nontraditional machining processes are chip-less material removal processes that involve the use of energy (rather than a cutting tool) for material cutting. All nontraditional machining processes can be classified into four categories: (1) mechanical-energy machining, (2) thermal-energy machining, (3) chemical machining, and, (4), electrochemical machining.

Mechanical-energy machining processes can be divided into four types: (1) ultrasonic machining, (2) abrasive jet machining (AJM), (3) abrasive water jet machining, and (4) water jet machining. Thermal energy machining processes are of the following four types: (1) electric discharge machining, (2) electric discharge wire cutting, (3) electron beam machining, and (4) laser beam machining. Additionally, the processes requiring electrochemical energy include electrochemical machining (ECM) and electrochemical deburring.

The nontraditional machining processes enable us to perform difficult-to-machine jobs (e.g., cutting of glass, high-strength alloys, etc.). A further description and mathematical modeling of nontraditional machining processes are given in Chapter 15.

1.2.3 Advanced Manufacturing Technologies

Advanced manufacturing technologies include computer-integrated manufacturing (CIM), additive manufacturing (AM), micromanufacturing, nanomanufacturing, and the like. CIM involves both computer-aided manufacturing (CAM) as well as all manufacturing engineering activities from computer-aided design (CAD) to shipment of the products to market. The description and mathematical modeling of CAM are given in Chapter 17.

AM covers a range of technologies that build three-dimensional (3D) objects by adding layer upon layer of material, whether the material is metallic, plastic, or ceramic (see Figure 1.3). The AM technologies include 3D printing, rapid prototyping (RP), direct digital manufacturing, layered manufacturing, and additive fabrication. AM essentially involves the use of a computer, 3D modeling software (CAD), processing equipment, and layering material. First, a CAD drawing is created. Then, the CAD data (CAD file) is read by the AM equipment, which lays downs or adds successive layers of liquid, powder, sheet material, or other, in a layer-upon-layer fashion to fabricate a 3D

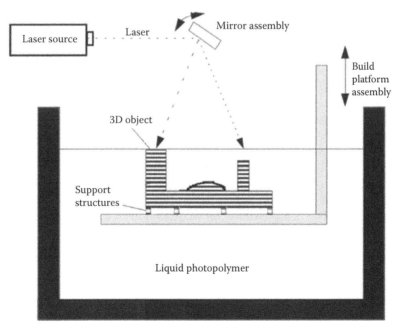

FIGURE 1.3
3D printing by stereolithography—an AM technology.

object (Figure 1.3). AM technologies are discussed in detail elsewhere (Stucker et al., 2014).

Micromanufacturing, also called microelectromechanical systems (MEMS), involves integrated circuits with the introduction of materials and products whose dimensions are in the micrometer range. Examples of the products made by MEMS include microsensors, compact discs, inkjet printing heads, and the like. *Nanomanufacturing*, also called nanoelectromechanical systems (NEMS), involves the introduction of materials and products whose feature sizes are in nanometer scale (Groover, 2014). Examples of products made by the NEMS include nanocoating for oil and gas industries, flat-screen TV monitors, ultrathin coatings for catalytic converters, cancer drugs, and the like. The Massachusetts Institute of Technology is actively involved in research and development in advanced manufacturing technologies.

1.3 Importance of Mathematical Modeling in Manufacturing

The engineering analysis/mathematical modeling in manufacturing is of great technological and social importance since it is almost impossible for a manufacturing/material process engineer to produce defect-free parts without solving the involved numerical problems. Consider metal casting, for example. In order to manufacture a defect-free sand-cast product, an engineer must properly design pattern, sprue, riser, and mold by using the relevant mathematical models and by solving the casting design problems. For instance, the pattern design calculations would ensure that the casting produced has the accurate size/dimensions (i.e., the casting would be neither undersized nor oversized). The sprue design calculations would ensure laminar flow of molten metal in the mold, thereby avoiding defect in the casting.

The faults in weld-design calculations may result in engineering failures, which are a threat to society. For example, failures of *Liberty Ships* during World War II are attributed to sites of stress concentrations in welds in the ships' structure, which was, of course, a serious fault in the weld design calculations. Another example is the significance of calculations for machining parameters in turning/drilling/milling. Here, the spindle rpm, N, must be computed on the basis of cutting speed, v, and the original diameter of the workpiece, D_o. The computations for material removal rate and cutting time are also industrially important so as to assure completion of the machining job well in time. The mathematical modeling of nontraditional machining processes (particularly AJM and ECM) is also very important.

Questions

1.1. Sketch a model of a manufacturing process.

1.2. List basic manufacturing processes.

1.3. "Product design is the job of marketing professionals." Is this statement true or false? Explain in either case.

1.4. Which manufacturing process is applicable to both metals and ceramics? Explain the process with the aid of a diagram.

1.5. Briefly explain the following manufacturing processes with the aid of diagrams: (a) metal casting and (b) metal forming.

1.6. Why is AM so named? Briefly explain the AM process with the aid of a diagram.

1.7. Differentiate between micromanufacturing and nanomanufacturing, giving examples of products manufactured by them.

References

Creese, R. 1999. *Manufacturing Processes and Materials*. Boca Raton, FL: CRC Press.

Dossett, J.L., Boyer, H.E. 2006. *Practical Heat Treating*. Materials Park, OH: ASM (International).

Groover, M.P. 2014. *Fundamentals of Modern Manufacturing*. New York: John Wiley and Sons.

Huda, Z. 2012. Reengineering of manufacturing process design for quality assurance in axle-hubs of a modern motor-car—A Case Study. *International Journal of Automotive Technology*, 13(7), 1113–1118.

Huda, Z. 2017. *Materials Processing for Engineering Manufacture*. Zurich, Switzerland: Trans Tech Publications.

James, W.B. 2015. Powder Metallurgy Methods and Applications. In *ASM Handbook, Vol. 7: Powder Metallurgy*. eds P. Samal and J. Newkirk, 9–19. Materials Park, OH: ASM (International).

Stucker, B., Rosen, D., Gibson, I. 2014. *Additive Manufacturing Technologies: 3D Printing, Rapid Prototyping, and Direct Digital Manufacturing*. New York: Springer-Verlag.

Youssef, H.A., El-Hofy, H.A., Ahmad, M.A. 2011. *Manufacturing Technology: Materials, Processes, and Equipment*. Boca Raton, FL: CRC Press/Taylor and Francis.

2

Metal Casting I—Casting Fundamentals

2.1 Melting of Metals in Foundry

Basics: Metal casting essentially involves introducing (pouring or forcing) molten metal into a mold so that the metal takes the shape of the mold cavity to produce the cast product (see Chapter 1, Figure 1.3). The supply of an adequate quantity and quality of molten metal is ensured by heating and melting metals in foundry furnaces. There are four types of furnaces generally used in a foundry: (1) crucible furnace, (2) cupola furnace, (3) electric arc furnace, and (4) induction furnace. Metal melting in a crucible furnace is explained in the following paragraph; the descriptions of the other melting furnaces are given elsewhere (Beeley, 2001; Tuttle, 2012; Huda, 2017).

A crucible furnace melts the metal without direct contact with fuel. A graphite container (crucible) is placed in a furnace and heated sufficiently to melt the metal charge (see Figure 2.1). Once the metal is melted, the crucible is lifted out of the furnace and used as a pouring ladle.

Mathematical Modeling: Total heat energy required for melting and pouring can be expressed as

$$Q = Q_1 + Q_2 + Q_3 \qquad (2.1)$$

where Q is the total heat energy required to heat the metal from room (ambient) temperature to pouring temperature, J; Q_1 is the heat energy to raise the temperature from starting temperature to melting point, J; Q_2 is the heat energy required to melt the metal, J; and Q_3 is the heat energy to raise the temperature of molten metal from melting temperature to pouring temperature (superheat), J.

The heat energy, Q_1, can be calculated by

$$Q_1 = mC_s \cdot (T_m - T_0) \qquad (2.2)$$

where m is the mass of metal being heated, g; C_s is the specific heat of solid metal, J/g-°C; T_m is the melting temperature of the metal, °C; T_0 is the starting (ambient) temperature, °C.

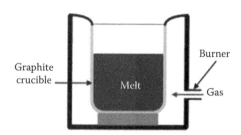

FIGURE 2.1
Crucible gas-fired furnace.

The heat energy, Q_2, can be determined by

$$Q_2 = mH_f \tag{2.3}$$

where H_f is the heat of fusion of the metal, J/g.

The heat energy, Q_3, can be computed by

$$Q_3 = mC_L \cdot (T_p - T_m) \tag{2.4}$$

where C_L is the specific heat of liquid metal, J/g-°C; and T_p is the pouring temperature, °C.

By using Equations 2.2 through 2.4, Equation 2.1 takes the following form:

$$Q = mC_s \cdot (T_m - T_0) + mH_f + mC_L \cdot (T_p - T_m) \tag{2.5}$$

The significance of Equation 2.5 is illustrated in Example 2.1.

2.2 Solidification of Metal

Basics: The solidification behavior of metal in a casting mold has a strong influence on the microstructure and mechanical properties of the casting (cast product). It is, therefore, important for the manufacturing/foundry engineer to properly design the solidification process. In particular, the rate of cooling or solidification strongly controls the microstructure of the metal. Rapid cooling (high solidification rate) would result in a fine-grained microstructure. Slow cooling (low solidification rate) of metal results in a coarse-grained microstructure. A high tensile strength is ensured by achieving a fine-grained microstructure.

The cooling behavior of a pure metal is usually illustrated by the cooling curve shown in Figure 2.2. The cooling curve indicates that a pure metal solidifies at a constant temperature that is equal to its freezing temperature, which is the same as its melting temperature. It is also evident in the cooling curve (Figure 2.2) that the *total solidification time*, t_s, is the sum of pouring time and

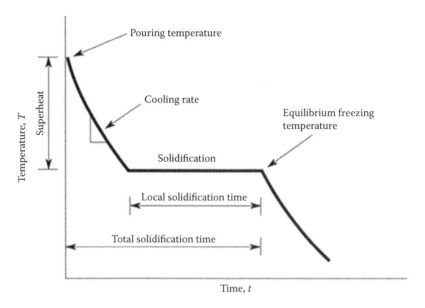

FIGURE 2.2
Cooling curve for a pure metal.

local solidification time. It means that the solidification time, t_s, is the time required for the casting from pouring to complete solidification. This time depends on the size and shape of the casting.

Mathematical Modeling: The solidification time, t_s, is related to the casting geometry by *Chvorinov's rule*, which is expressed by the following formula (Campbell, 2011):

$$t_s = B(V/A)^n \tag{2.6}$$

where t_s is the total solidification time, min (see Figure 2.2); B is the mold constant or solidification constant, min/cm^2; n is a constant ($n = 1.5$ to 2); V is the volume of casting, cm^3; and A is the surface area through which heat is extracted, cm^2. In most cases, the value of the constant $n = 2$. Hence, *Chvorinov's law* (Equation 2.6) may also be expressed as

$$t_s = B(V/A)^2 \tag{2.7}$$

It is important to note that Equation 2.7 has a limitation; it is applicable for castings that involve one-dimensional heat transfer (e.g., plates). Jelínek and Elbel (2010) have overcome this limitation by generalizing this relationship to include heat transfer on concave mold surfaces curved inward (e.g., a cylinder, a ball, etc.), which is expressed as follows:

$$t_s = \varepsilon \cdot B \cdot (V/A)^2 \tag{2.8}$$

where ε is a shape coefficient. It may be noted that for a plate form casting, $\varepsilon = 1$, so Equation 2.8 would simplify to Equation 2.7; however, for other forms (cylinder, ball, etc.), $\varepsilon < 1$ (Jelínek and Elbel, 2010).

The significances of Equations 2.6 through 2.8 are illustrated in Examples 2.2 through 2.4.

2.3 Metal Casting Processes

Manufacturing a defect-free casting is a challenging job since it requires strict process control and resource availability. There are six basic requirements for metal casting: (1) a mold must be produced with a *mold cavity*, having the desired geometrical design of the casting (solidified metal/product); (2) a *melting furnace* must be capable of providing molten metal at the right temperature in the desired quantity and quality; (3) a *pouring facility* must be available to admit the molten metal into the mold to fill the cavity; (4) the *solidification process* should be properly designed to ensure the desired cooling rate; (5) a proper *mold removal* technique must be devised; and (6) tools should be available for cleaning, finishing, and inspecting the cast product. There are a variety of different types of metal casting processes. Depending on the type of mold used, all casting processes may be classified into two groups: (1) expendable mold casting and (2) permanent mold casting. Expendable mold casting processes involve the use of temporary, nonreusable molds. Permanent mold casting processes involve a permanent mold that is reusable again and again.

2.3.1 Expendable Mold Casting Processes

There are four main types of expendable mold casting processes: (1) sand-casting, (2) shell mold casting, (3) lost foam casting, and (4) investment casting.

The expendable mold casting processes are briefly explained in the following sections.

2.3.1.1 Sand Casting

Basics: Sand-casting is the most common and the most economical metal casting process. It is so named because the mold is made from sand. Figure 2.3 illustrates the steps in sand-casting.

Some components are designed to have a hole; this casting design requires the use of a *core* in the mold. A *core* is actually a separate part of the mold, made of baked sand, which is used to create holes and various shaped cavities in the castings (see Figure 2.4). As the mold fills with the molten metal, it is exposed to high metallostatic forces, which tend to displace the mold sections and cores. It is important for a foundryman to determine the metallostatic and

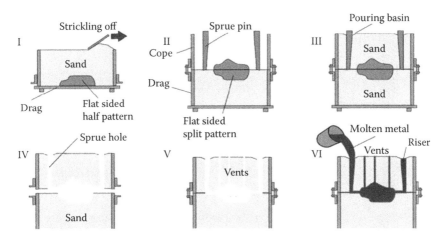

FIGURE 2.3
The steps in sand molding and casting.

buoyancy forces. In order to prevent core displacement, the core is held in position by the use of chaplets (Figure 2.4).

Mathematical Modeling: It is explained in the preceding paragraph that the metallostatic force, gravity, and buoyancy forces act on the mold and core during the sand-casting process. The upward metallostatic force acting on a flat mold surface is given by

$$F_m = 9.81\rho_m h A_p \tag{2.9}$$

where F_m is the metallostatic force, N; ρ_m is the density of the metal, kg/m^3; h is the head of metal, m; and A = projection area of the casting on the parting plane, m^2.

Figure 2.5 shows a sand-casting mold, which indicates that head of metal = $h = H - a$.

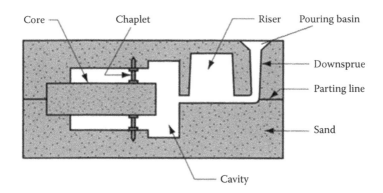

FIGURE 2.4
Sand mold showing core and chaplets.

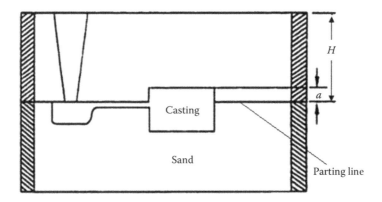

FIGURE 2.5
Head of metal ($h = H - a$) in a sand mold.

There is a gravity force (i.e., weight of the upper flask [cope] and the sand in the cope); this gravity force can be calculated by the following relationship:

$$W_s = V_s \rho_s g \tag{2.10}$$

where W_s is the weight of sand along with the cope, N; V_s is the volume of sand in the cope, m³; ρ_s is the density of sand, kg/m³; and $g = 9.81$ m/s².

In ferrous castings, the metallostatic force, F_m, is significantly higher than the gravity, W_s. Hence, it is important to either tightly clamp the cope with drag or use weights to prevent lifting of the cope (upper flask); this required weight can be calculated as follows:

$$W = F_m - W_s \tag{2.11}$$

where W is the weight required to prevent lifting of the cope, N (see Example 2.5).

In addition to metallostatic force, there may also be buoyancy force that tends to lift the core upward (see Figure 2.4); this force can be calculated as follows:

$$F_b = W_m - W_c = (m_1)g - (m_2)g \tag{2.12}$$

where F_b is buoyancy force, N; W_m is weight of the molten metal displaced by a core, N; W_c is the weight of core, N; $g = 9.81$ m/s²; m_1 is the mass of metal, kg; and m_2 is the mass of core, kg.

Density is defined as the ratio of mass to volume; this relationship can be written as

$$m_1 = V\rho_m \tag{2.13}$$

$$m_2 = V\rho_c \tag{2.14}$$

TABLE 2.1

Densities of Various Materials at 25°C

Material	Density (g/cm^3)
Steel	7.8
Cast iron	7.16
Aluminum	2.7
Sand	1.6
Copper	8.9

where V is the volume of the core (= volume of metal displaced), m^3; ρ_m is the density of metal, kg/m^3; and ρ_c is the density of the core (sand), kg/m^3.

By using Equations 2.12 and 2.13, we can rewrite Equation 2.12 as follows:

$$F_b = gV\ (\rho_m - \rho_c) \qquad (2.15)$$

The significance of Equations 2.12 through 2.15 is illustrated in Examples 2.6 through 2.8.

The values of densities of various materials are presented in Table 2.1.

2.3.1.2 Shell Mold Casting

Shell molding, or shell mold casting involves the formation of a shell mold that is made from resin-covered sand. Once the shell molds are created, they are then assembled and the metal poured into the mold cavity. On completion of solidification, the part is removed. The five steps in shell molding are illustrated in Figure 2.6.

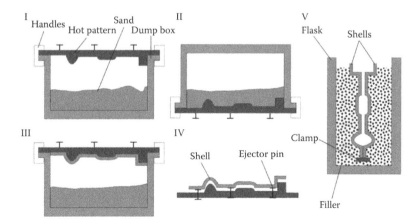

FIGURE 2.6
The five steps in shell molding. (I) The heated pattern is clamped to a dump box. (II) The dump box is inverted, allowing this sand-resin mixture to coat the pattern. (III) The heated pattern partially cures the mixture, which now forms a shell around the pattern. (IV) The shell is ejected from the pattern. (V) Mold is assembled.

2.3.1.3 Lost Foam Casting

In the *lost foam casting* process, a pattern of polystyrene is first coated with a refractory compound. Then the foam pattern is placed in a mold box, and sand is compacted (by ramming) around the pattern. Once the mold is ready, molten metal is poured into the portion of the pattern that forms the pouring cup and sprue. As the hot metal enters the mold (pattern), the polystyrene foam is vaporized ahead of the advancing liquid, thereby allowing the resulting mold cavity to be filled. On solidification, the part is removed.

2.3.1.4 Investment Casting

Investment casting, also called *lost-wax casting*, is an expandable mold casting process that involves the use of a wax pattern that melts out (lost) during casting. In this process, first a wax pattern is coated with a refractory ceramic material that is hardened with time, and the internal geometry of the ceramic coating takes the shape of the casting (mold cavity). The mold is then inverted and heated so that wax is melted out of the cavity. Then the mold is held upright and molten metal is poured into the cavity formed due to melting of the wax. On completion of solidification within the ceramic mold, the metal casting is removed.

2.3.2 Permanent Mold Casting Processes

There are four main types of permanent mold casting processes: (a) gravity die casting (GDC), (b) low-pressure permanent mold casting (LPPMC), (c) die casting, and (d) centrifugal casting.

The permanent mold casting processes are described in the following subsections.

2.3.2.1 Gravity Die Casting

GDC involves the use of a permanent mold (made of steel alloy) and gravity to fill the mold with the molten metal. GDC is well suited for high-volume production of zinc and copper alloy products. In GDC, first a heated mold (die) is coated with a die release agent. Then molten metal is poured into channels in the die tool to allow the liquid metal to fill all the extremities of the die cavity. On cooling, the die is opened and the part (casting) is removed.

2.3.2.2 Low-Pressure Permanent Mold Casting

In the LPPMC process, a gas at a low pressure (~60 kPa) is used to push the molten metal into the cavity in a permanent mold. The low pressure is applied to the top of the pool of molten metal, which forces the liquid up a refractory tube, and finally into the bottom of the mold.

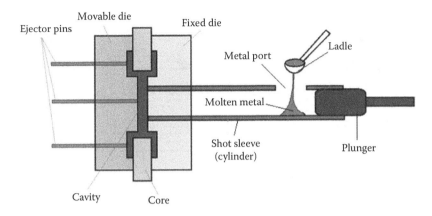

FIGURE 2.7
Cold chamber die casting process.

2.3.2.3 Die Casting

The *die casting* process is also called the *high-pressure die casting* process because it involves forcing molten metal under high pressure into a die cavity. Like GDC, the die is made by using two hardened tool-steel die halves, which have been machined to create a die cavity as per the geometric requirements of the cast product. Most die castings (products) are made from low melting–temperature nonferrous metals (e.g., zinc, magnesium, aluminum, lead, etc.). There are two main types of die casting machines: (a) hot-chamber die casting machine and (b) cold-chamber die casting machine. A hot-chamber die casting machine, or *gooseneck machine*, relies on a pool of molten metal to feed the die. In the cold-chamber die casting process, the molten metal reservoir is separated from the injection system. In the cold-chamber machine, the molten metal is poured from the ladle into a steel shot sleeve (see Figure 2.7). The shot sleeve is typically at a temperature in the range of 200°C–300°C. Then a hydraulically operated plunger forces the molten metal at high pressure into the die cavity resulting in rapid die filling. The steel die dissipates the latent heat of fusion. During solidification, the metal in the die cavity is kept at high pressure by the plunger to feed the solidification shrinkage. On cooling, the die is opened, and the casting is ejected (Andresen, 2004).

2.3.2.4 Centrifugal Casting

Basics: In *centrifugal casting,* a permanent mold is rotated continuously about its axis at high speeds as the molten metal is poured into the mold (see Figure 2.8). The rotational speed of the mold ranges from 300 to 3000 rpm. In a true centrifugal casting manufacture, the molds used are round, and are typically made of iron, steel, or graphite (Figure 2.8). As the mold rotates, the molten metal is centrifugally thrown toward the inside mold wall, where it cools and solidifies. Owing to the chilling effect against the mold surface, the casting

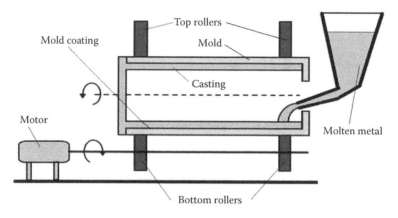

FIGURE 2.8
Centrifugal casting process.

has a fine-grained microstructure. The impurities and nonmetallic inclusions in the molten metal are also centrifugally thrown to the inside mold wall (outer skin of the casting), which can be machined away (Campbell, 2011).

Mathematical Modeling: Owing to the rotation of the mold, centrifugal force acts on the molten metal, which is given by

$$F = (mv^2)/R \tag{2.16}$$

where F is the centrifugal force, N; m is the mass of molten metal, kg; v is the tangential velocity of the metal, m/s; and R is the internal radius of the mold, m.

In centrifugal casting, the ratio of the centrifugal force experienced by the rotating cast metal to its weight, is called G-factor, G_f. It means the following:

$$G_f = F/\text{Weight} = F/mg \tag{2.17}$$

The linear speed of the molten metal, v, is related to the rotational speed of the mold, N, by

$$v = 2\pi RN \tag{2.18}$$

where v is the linear speed, m/s; and N is the rotational speed, rev/s.

Since rotational speed of the mold, N, is expressed in rpm, Equation 2.18 can be rewritten as

$$v = 2\pi RN/60 = \pi RN/30 \tag{2.19}$$

where N is expressed in rev/min.

By combination of Equations 2.16 through 2.19, we can work out the following:

$$G_f = F/mg = [(mv^2)/R]/mg = v^2/Rg = (\pi RN/30)^2/Rg$$

$$N = (30/\pi)[(2G_f g)/D)^{1/2}] \tag{2.20}$$

where N is the rotational speed of the mold, rev/min; G_f is the G-factor; $g = 9.8 \, \text{m/s}^2$ and $D =$ inside diameter of mold, $m =$ outside diameter of the casting, m.

In general, the values of the G-factor are in the range of 60–68.

The significance of Equations 2.16 through 2.20 is illustrated in Examples 2.11 and 2.12.

2.4 Problems and Solutions—Examples in Metal Casting

EXAMPLE 2.1: COMPUTING HEAT ENERGY REQUIRED TO HEAT A METAL TO POURING TEMPERATURE

A 4 kg commercially pure aluminum is to be heated and melted for casting. Calculate the total heat energy required to heat the metal from the ambient temperature (25°C) to the pouring temperature (750°C).

Solution

Mass $= m = 4000$ g; starting temperature $= T_0 = 25°C$; pouring temperature $= T_p = 750°C$.

For aluminum, $T_m = 660.3°C$, $C_s = 0.91 \, \text{J/g.°C}$, $C_L = 1.18 \, \text{J/g.°C}$, $H_f = 398 \, \text{J/g}$.

By using Equation 2.5,

$$Q = mC_s \cdot (T_m - T_0) + mH_f + mC_L \cdot (T_p - T_m)$$
$$= 2{,}312{,}492 + 1{,}592{,}000 + 423{,}384 = 4{,}327{,}876 \, \text{J}$$

Hence, the total heat energy required $= 4.33$ MJ.

EXAMPLE 2.2: CALCULATING SOLIDIFICATION TIME FOR A CASTING

A rectangular steel product is to be produced by the sand-casting process. The dimensions of the product are shown in Figure E2.2. Calculate the solidification time, if the mold constant of Chvorinov's rule is $4.0 \, \text{min/cm}^2$.

Solution

The value of n should be in the range of 1.5–2. By assuming, $n = 2$,

$$\text{Surface area} = A = 2 \, [(16 \times 8.5) + (16 \times 3.5) + (8.5 \times 3.5)] = 443 \, \text{cm}^2.$$

$$\text{Volume} = V = 16 \times 8.5 \times 3.5 = 476 \, \text{cm}^3.$$

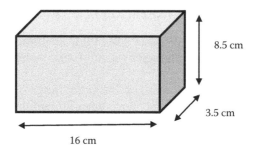

FIGURE E2.2
Dimensions of the steel casting.

By using Equation 2.6,

$$t_s = B\left(\frac{V}{A}\right)^n = 4.0\left(\frac{476}{443}\right)^2 = 4.6 \text{ min}$$

The solidification time $= 4.6$ min.

EXAMPLE 2.3: CALCULATING THE MOLD CONSTANT

A metal plate having dimensions (3 cm × 5 cm × 10 cm) was produced by sand-casting. Taking $n = 2$, calculate the mold constant if the casting solidification time was 3 minutes.

Solution

Solidification time $= t_s = 3$ min; $B = ?$; Volume $= V = 3$ cm × 5 cm × 10 cm $= 150$ cm³.

Surface area through which heat is extracted $= A = [2(3 \times 5)] + [2(5 \times 10)] + [2(10 \times 3)] = 190$ cm².

By the application of *Chvorinov's rule* (Equation 2.6),

$$t_s = B(V/A)^n$$

$$3 = B(150/190)^2 \text{ or } B = 4.81 \text{ min/cm}^2$$

Hence, the mold constant $= 4.81$ min/cm².

EXAMPLE 2.4: CALCULATING THE SOLIDIFICATION TIME FOR TWO DIFFERENT CASTINGS

A metal plate casting having dimensions (10 cm × 5 cm × 3 cm) solidifies in 3 minutes. Calculate the solidification time for another casting having dimensions (13 cm × 6 cm × 4 cm) if the same metal is poured under the same conditions.

Solution

Casting 1 having dimensions (10 cm × 5 cm × 3 cm)

$t_s = 3$ min, $V = 3$ cm × 5 cm × 10 cm = 150 cm³; $A = 190$ cm²

By using Equation 2.7,

$$t_s = B(V/A)^2,$$

$$3 = B(150/190)^2$$

Hence, the mold constant $= B = 4.81$ min/cm².

Casting 2 having dimensions (13 cm × 6 cm × 4 cm),

$$V = 13 \text{ cm} \times 6 \text{ cm} \times 4 \text{ cm} = 312 \text{ cm}^3$$

$$A = [2(13 \times 6)] + [2(6 \times 4)] + [2(4 \times 13)] = 308 \text{ cm}^2$$

Since pouring conditions are the same, the mold constant is the same (i.e., $B = 4.81$ min/cm²).

By using Equation 2.7,

$$t_s = B \ (V/A)^2 = 4.81(312/308)^2 = 4.93 \text{ min}$$

Hence, the solidification time for the casting with (13 cm × 6 cm × 4 cm) dimensions $= 4.93$ min.

EXAMPLE 2.5: CALCULATING THE SOLIDIFICATION TIME FOR A SPHERICAL METAL CASTING

A metal solid sphere having 10 cm diameter is to be produced by casting. The shape coefficient is 0.9, and the mold constant is 3.8 min/cm². Calculate the solidification time for the casting.

Solution

Mold constant $= B = 3.8$ min/cm², shape coefficient $= \varepsilon = 0.9$, radius $= r = 5$ cm.

Volume of sphere $= V = (4/3)\pi r^3 = (4/3)\pi(5)^3 = 523.6$ cm³.

Surface area of sphere $= A = 4\pi r^2 = 4\pi(5)^2 = 314.16$ cm².

By using Equation 2.8,

$$t_s = \varepsilon \cdot B \cdot (V/A)^2$$

$$t_s = 0.9 \times 3.8 \times (523.5/314.16)^2 = 0.9 \times 3.8 \times 2.77 = 9.5 \text{ min}$$

EXAMPLE 2.6: CALCULATING THE METALLOSTATIC FORCE AND THE REQUIRED WEIGHT FOR A SAND MOLD

A cast iron component (shown in Figure E2.6a) is required to be cast by sand-casting. If the flask dimensions are 320 × 320 × 200 mm³, calculate

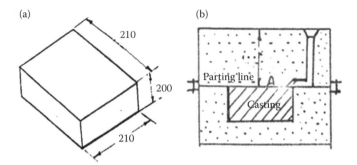

FIGURE E2.6
(a) Cast iron component (all dimensions in mm). (b) Mold showing casting located fully below the parting line.

(a) the metallostatic force and (b) the required weight that should be used to prevent lifting of the upper flask (cope).

Solution

If the mold cavity is formed fully in the drag of the mold (fully below the parting line), the sand mold may be shown as in Figure E2.6b.

(a) Density of cast iron $= \rho_m = 7.16\,g/cm^3 = 7160\,kg/m^3$ (see Table 2.1).

Head of metal $= h = H - a = 200 - 0 = 200\,mm = 0.2\,m$ (casting fully below the parting line).

Projection area of the casting $= A_p = 210\,mm \times 210\,mm = 0.21\,m \times 0.21\,m = 0.044\,m^2$.

By using Equation 2.9,

$$F_m = 9.81 \rho_m h A_p = 9.81 \times 7160 \times 0.2 \times 0.044 = 619.5\,N$$

Hence, metallostatic force $= 620\,N$.

(b) $F_m = 620\,N$, Volume of sand in the cope $= V_s = 320\,mm \times 320\,mm \times 200\,mm = 0.0205\,m^3$.

Density of sand $= \rho_s = 1.67\,g/cm^3 = 1670\,kg/m^3$.

The gravity force (i.e., weight of the sand [and cope]) can be found by using Equation 2.10:

$$W_s = V_s \rho_s g = 0.0205 \times 1670 \times 9.81 = 335.5\,N$$

Now, the required weight can be calculated by using Equation 2.11 as follows:

$$W = F_m - W_s = 620 - 335.5 = 284.5\,N \sim 285\,N$$

$$W = mg,\ 285 = m \times 9.81,\ m = 29\,kgf$$

Hence, 30 kgf weights should be used to prevent lifting of the cope (against buoyancy).

EXAMPLE 2.7: CALCULATING THE BUOYANCY FORCE ON THE CORE WHEN THE CORE VOLUME IS GIVEN

A sand core located inside a mold cavity has a volume of 2450 cm^3. It is used in the casting of a cast iron component. Determine the buoyancy force on the core.

Solution

Volume of metal displaced by the core $= 2450 \text{ cm}^3 = 0.00245 \text{ m}^3$.

By reference to Table 2.1,

Density of metal (cast iron) $= \rho_m = 7.16 \text{ g/cm}^3 = 7160 \text{ kg/m}^3$.

Density of sand $= \rho_c = 1.6 \text{ g/cm}^3 = 1600 \text{ kg/m}^3$.

By using Equation 2.15

$$\text{Buoyancy force on the core} = F_b = W_m - W_c = g \, V \, (\rho_m - \rho_c),$$

$$F_b = 9.81 \times 0.00245(7160 - 1600) = 133.6 \text{ N}$$

Hence, buoyancy force on the core $= 133.6 \text{ N}$.

EXAMPLE 2.8: CALCULATING THE WEIGHT NEEDED TO SUPPORT THE CORE AGAINST BUOYANCY FORCE

An aluminum alloy casting is to be made in a sand mold using a sand core weighing 15 kgf. What weight is needed to prevent core lift during pouring? Density of the alloy $= 2.79 \text{ g/cm}^3$.

Solution

By reference to Table 2.1,

Density of metal $= \rho_m = 2.79 \text{ g/cm}^3 = 0.00279 \text{ kg/cm}^3$.

Density of core (sand) $= \rho_c = 1.6 \text{ g/cm}^3 = 0.0016 \text{ kg/cm}^3$.

Mass of core $= m_2 = 15 \text{ kg}$.

Volume of core $= V = m_2/\rho_c = (15/0.0016) = 9375 \text{ cm}^3$.

By using Equation 2.15,

$$F_b = W_m - W_c = gV(\rho_m - \rho_c)$$
$$= 9.81 \times 9375 \times (0.00279 - 0.0016) = 109.44 \text{ N}$$

The buoyancy force ($F_b = 109.44 \text{ N}$) acts upward, so an equal downward force (weight $= W = 109.44 \text{ N}$) must be used to prevent the core displacement:

$$m = W/g = (109.44)/(9.81) = 11.15 \text{ kgf}$$

Hence, a weight of at least 12 kgf must be used to prevent the core displacement.

EXAMPLE 2.9: CALCULATING MASS OF THE CASTING MADE BY SAND-CASTING USING A CORE

A sand core used to form the internal surfaces of a steel casting experiences a buoyancy force of 20 kgf. The volume of the mold cavity forming the outside surface of the casting is 4800 cm³. What is the mass of the final casting?

Solution

Density of steel $= 0.0078\,\text{kg/cm}^3$ (see Table 2.1).

Density of sand $= 0.0016\,\text{kg/cm}^3$ (see Table 2.1).

Volume of outside surface of casting $= 4800\,\text{cm}^3$.

Buoyancy force on core $= F_b = mg = 20 \times 9.81 = 196.2\,\text{N}$.

By using Equation 2.12,

$$F_b = W_m - W_c = (m_m - m_c) \times g$$

$m_m =$ density of metal $\times V_m = 0.0078\,V_m$

$m_c =$ density of core $\times V_c = 0.0016 \times V_c$

$V_m =$ Volume of metal displaced by core $= V_c = V$

By substituting the above values in Equation 2.12, we get

$$F_b = (0.0078\,\text{V} - 0.0016\,\text{V}) \times g$$

$$196.2 = (0.0078\,\text{V} - 0.0016\,\text{V}) \times 9.81$$

$$V = 3225.8\,\text{cm}^3$$

Hence, volume of core $= 3225.8\,\text{cm}^3$.

Volume of casting $=$ Volume of outside surface of cavity – Volume of core.

$$\text{Volume of casting} = 4800 – 3225.8 = 1574\,\text{cm}^3.$$

Mass of steel casting $=$ density $\times$ volume $= 0.0078 \times 1574 = 12.27\,\text{kg}$.

EXAMPLE 2.10: CALCULATING THE MASS OF METAL TO BE POURED FOR TRUE CENTRIFUGAL CASTING

A steel hollow cylinder having OD $= 10\,\text{mm}$, ID $= 3\,\text{mm}$, and L $= 150\,\text{cm}$, is required to be manufactured by true centrifugal casting. Calculate the mass of steel required to be poured for the casting. Ignore all allowances.

Solution

OD $= 10\,\text{mm} = 1\,\text{cm}$, ID $= 3\,\text{mm} = 0.3\,\text{cm}$, L $= 150\,\text{cm}$.

Volume of metal in the cylinder $= V =$ Outer volume $-$ Inner volume $= V_o - V_i$.

$$V = (\pi/4)L[(OD)^2 - (ID)^2] = (0.785) \times 150 \times [(1)^2 - (0.3)^2] = 107.2 \text{ cm}^3$$

By reference to Table 2.1, density of steel $= \rho = 7.8 \text{ g/cm}^3$.

$$\text{Mass} = \rho V = (7.8)(107.2) = 836.1 \text{ g}$$

Mass of steel required to be poured $= 0.84$ kg.

EXAMPLE 2.11: DETERMINING THE ROTATIONAL SPEED OF MOLD IN CENTRIFUGAL CASTING

By using the data in Example 2.10, calculate the rotational speed of the mold if G-factor is 66.

Solution

Outside diameter of casting $= D = 1$ cm $= 0.01$ m, $G_f = 66$, $N = ?$ rpm.

By using Equation 2.20, we get

$$N = (30/\pi) [(2G_f g)/D)^{1/2}] = (9.55) [(2 \times 66 \times 9.81)/0.01]^{1/2}$$
$$= (9.55) \times (359.4) = 3436 \text{ rpm}$$

Hence, the rotational speed of the mold $= 3436$ rev/min.

EXAMPLE 2.12: CALCULATING THE CENTRIFUGAL FORCE IN CENTRIFUGAL CASTING

By using the data in Examples 2.10 and 2.11, calculate the centrifugal force acting on the molten metal during the centrifugal casting process.

Solution

Mass of steel $= m = 0.836$ kg (see Example 2.10), $N = 3436$ rev/min (see Example 2.11).

By using Equation 2.19,

$$v = (\pi RN)/30 = (\pi \times 5 \times 3436)/30 \text{ mm/s} = 1799 \text{ mm/s} = 1.8 \text{ m/s}$$

$$v^2 = 3.24, R = 5 \text{ mm} = 0.005 \text{ m}$$

By using Equation 2.16,

$$F = (mv^2)/R = (0.836 \times 3.24)/0.005 = 541 \text{ N}$$

Centrifugal force $= 541$ N.

Questions and Problems

2.1 Define the following terms: (a) metal casting, (b) foundry, and (c) sand core.

2.2 What are the basic requirements in metal casting?

2.3 Distinguish between expendable mold casting and permanent mold casting processes giving at least three examples for each group of casting process.

2.4 Explain the die casting process with the aid of a sketch.

2.5 Copper scrap weighing 3 kgf is to be heated and melted for casting. Calculate the total heat energy required to heat the metal from the ambient temperature ($25°C$) to the pouring temperature ($1150°C$).

2.6 A metal plate having dimensions ($13 \text{ cm} \times 8 \text{ cm} \times 4 \text{ cm}$) is to be produced by sand-casting. Calculate the mold constant if the casting solidifies in 4 minutes.

2.7 By using the mold constant value in Problem (2.6), calculate the solidification time of a casting having dimensions ($15 \text{ cm} \times 9 \text{ cm} \times 3 \text{ cm}$) if the casting is made of the same metal and poured under the same conditions.

2.8 A metallic cylinder 25 cm long having 8 cm diameter is to be produced by sand-casting. The shape coefficient is 0.87, and the mold constant is 3.9 min/cm^2. Calculate the solidification time for the casting.

2.9 An aluminum rectangular component having dimensions ($180 \text{ mm} \times 160 \text{ mm} \times 100 \text{ mm}$) is required to be produced by sand-casting. If the flask dimensions are ($330 \times 330 \times 200 \text{ mm}^3$), calculate (a) the metallostatic force and (b) the required weight that should be used to prevent lifting of the upper flask (cope).

2.10 A sand core located inside a mold cavity has a volume of 2000 cm^3. It is used in the casting of a steel component. Determine the buoyancy force on the core.

2.11 A cast iron casting is to be made in a sand mold using a sand core weighing 15 kgf. What weight is needed to prevent core lift during pouring?

2.12 A sand core used to form the internal surfaces of an aluminum casting experiences a buoyancy force of 14 kgf. The volume of the mold cavity forming the outside surface of the casting is 4500 cm^3. What is the mass of the final casting?

2.13 A copper hollow cylinder ($L = 200 \text{ cm}$, $OD = 15 \text{ cm}$, $ID = 10 \text{ cm}$) is required to be manufactured by true centrifugal casting. Calculate (a) the mass of copper metal required to be poured for the casting, ignoring all allowances; (b) rotational speed of the mold, rev/min; and (c) centrifugal force on the molten metal.

References

Andresen, W. 2004. *Die Cast Engineering: A Hydraulic, Thermal, and Mechanical Process.* Boca Raton, FL: CRC Press/Taylor and Francis.

Beeley, P.R. 2001. *Foundry Technology.* Oxford: Butterworth-Heinemann.

Campbell, J. 2011. *Complete Casting Handbook: Metal Casting Processes, Techniques and Design.* Oxford, UK: Butterworth-Heinemann/Elsevier Science.

Huda, Z. 2017. *Materials Processing for Engineering Manufacture.* Switzerland: Trans Tech Publications.

Jelínek, P., Elbel, T. 2010. Chvorinov's rule and determination of coefficient of heat accumulation of molds with non-quartz base sands. *Archives of Foundry Engineering*, 10(4), 77–82.

Tuttle, R.B. 2012. *Foundry Engineering: The Metallurgy and Design of Castings.* Colorado Springs, CO: Create Space Independent Publishing Platform.

3

Metal Casting II—Casting Design

3.1 Casting Defects

The defects in a sand-casting may be the result of one or more of the following factors: (1) improper metal/alloy composition, (2) improper melting practice, (3) improper pattern design, (4) improper pouring practice (including bad sprue design), (5) improper mold and core construction, (6) improper selection of molding and core-making materials, and (7) improper gating system (including bad riser design). All casting defects may be classified into three categories: (1) visible defects, (2) surface defects, and (3) internal defects (Goodrich, 2008; Campbell, 2011).

3.1.1 Visible Casting Defects

The visible casting defects include: hot tears, shrinkage cavity, swell, misrun, wash, and rat tail.

Hot Tear: These are cracks that appear in the form of irregular crevices with a dark oxidized fracture surface. The occurrence of a hot tear defect may be prevented by using: a relatively lower temperature of casting metal, ensuring reduced metal contraction, and correct design of the gating system and casting on the whole.

Shrinkage Cavity: A depression or an internal void in a casting resulting from the volume contraction during solidification is called *shrinkage cavity.* A proper riser design would ensure adequate molten metal supply to the mold cavity, thereby eliminating the shrinkage cavity defect.

Swell: A slight, smooth bulge usually found on vertical faces of castings is called a *swell.* The main cause of swell is the liquid metal pressure, which may be due to low strength of mold because of too high a water content or when the mold is not rammed sufficiently.

Misrun: The incomplete filling of a mold cavity is called *misrun.* The causes of *misrun* include too low melt temperature, inadequate metal supply, improperly designed gates, or too large length-to-thickness ratio of the casting.

Wash: A *wash* is a low projection on the drag face of a casting that extends along the surface, decreasing in height as it extends from one side of the casting to the other end. This defect can be prevented in bottom gating castings by ensuring high hot strength in the molding sand and a sufficient quantity of molten metal to flow through one gate into the mold cavity.

Rat tail: The small ridge of sand that extends into the mold cavity as a result of sand expansion can create a line on the surface of the casting is called a *rat tail*. It is caused by thermal expansion of the molding sand. A *rat tail* defect can be prevented by mixing wood flour to the sand.

3.1.2 Surface Defects

The surface defects that commonly occur include blow, scab, drop, penetration, and buckle.

Blow: Blow or *sand blow* is a relatively large cavity produced by gases that displace molten metal during solidification. This defect can be prevented by adequate venting of the molding sand and/or reducing the moisture content in the sand.

Scab: The hard, thick, and rough layer on the surface of the casting is called *scab*. This defect can be prevented by using a coarser sand of higher permeability, and by ensuring even ramming.

Drop: An irregularly shaped projection on the cope surface of a casting is called *drop*. It is caused by the breaking away of a part of mold sand. *Drop* can be prevented by (1) harder ramming of the mold sand, (2) using stronger molding sand, (3) proper functioning of molding equipment, or (4) avoiding strong jolts and strikes at the flask when assembling the mold.

Penetration: A strong crust of fused sand on the surface of a casting is called *penetration*. This defect can be avoided by (1) sufficient refractoriness of molding materials, (2) a smaller content of impurities, (3) adequate mold packing, or (4) using a better quality of mold washes.

Buckle: A buckle is a long V-shaped depression that occurs on the surface of flat castings. It results due to thermal expansion of sand caused by the heat of the metal. This defect can be prevented (1) by ensuring a good hot strength of the molding sand, (2) by using a better casting design that avoids too large a flat surface in the mold cavity, or (3) by mixing cereal to sand.

3.1.3 Internal Casting Defects

Blow Holes: Well-rounded cavities having a smooth surface in the body of a casting are called *blow holes, gas holes,* or *gas cavities*. These defects can be prevented by (1) reducing moisture content (in the case of green sand molds), moisture on chills, chaplets, or metal inserts; (2) ensuring adequate gas permeability of the molding sand; (3) adequately venting the mold; (4) sufficiently drying the mold and cores; (5) using a higher pouring temperature, and/or (6) correctly feeding the casting.

Pin Holes: The small gas holes either at the surface or just below the surface of the casting are called pin holes. This defect can be avoided by proper degassing of the molten metal/alloy prior to its pouring in the mold.

3.2 Pattern Design

3.2.1 Pattern and Pattern Design Considerations

A pattern is a model for the object to be cast (i.e., a pattern makes an impression on the mold into which liquid metal is poured); the metal solidifies in the shape of the original pattern. Patterns may be made of wood, metal, or plastic. The pattern design mainly involves the provision of various *pattern allowances*. A pattern should be slightly oversized to allow for shrinkage, draft, and machining. Additionally, sharp corners should be avoided by allowing radius of curvature in the pattern. Another important consideration in the pattern design is that a straight parting line should be made at mid-height of the casting (see Figure 3.1).

Table 3.1 presents machining allowances for a wooden pattern for various metals and pattern sizes. The typical values of linear shrinkage for various cast metals are listed in Table 3.2.

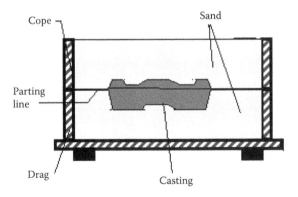

FIGURE 3.1
Sand mold showing parting line and gating system.

3.2.2 Mathematical Modeling of Pattern Design/Allowances

The procedure for calculating various pattern allowances for a wooden pattern, for a metallic gear (Figure 3.2), is illustrated in Table 3.3; its significance is given in Examples 3.1 and 3.2.

TABLE 3.1

Pattern (Machining) Allowances for a Wooden Pattern

Pattern Size (mm)	Bore	Surface	Cope Side
For cast steels			
Up to 152	3.2	3.2	6.4
152–305	6.4	4.8	6.4
305–510	6.4	6.4	7.9
510–915	7.1	6.4	9.6
915–1524	7.9	6.4	12.7
For cast irons			
Up to 152	3.2	2.4	4.8
152–305	3.2	3.2	6.4
305–510	4.8	4.0	6.4
510–915	6.4	4.8	6.4
915–1524	7.9	4.8	7.9
For nonferrous alloys			
Up to 76	1.6	1.6	1.6
76–152	2.4	1.6	2.4
152–305	2.4	1.6	3.2
305–510	3.2	2.4	3.2
510–915	3.2	3.2	4.0
915–1524	4.0	3.2	4.8

TABLE 3.2

Linear Shrinkage Values for Different Casting Metals

Metal	Linear Shrinkage	Metal	Linear Shrinkage	Metal	Linear Shrinkage
Aluminum alloys	1.3%	Magnesium	2.1%	Steel, chrome	2.1%
Brass, yellow	1.3%–1.6%	Magnesium alloy	1.6%	Tin	2.1%
Cast iron, gray	0.8%–1.3%	Nickel	2.1%	Zinc	2.6%
Cast iron, white	2.1%	Steel, carbon	1.6%–2.1%		

Source: Groover, M.P. 2014. *Fundamentals of Modern Manufacturing.* New York: John Wiley and Sons.

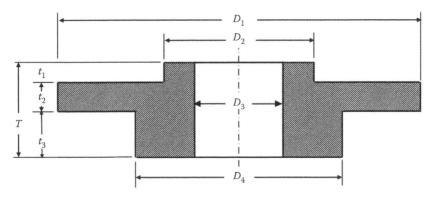

FIGURE 3.2
Generalized dimensions of a gear for pattern allowance calculation.

TABLE 3.3

Pattern Allowance Calculations for a Gear

Nominal Dimension	Machining Allowance	Shrinkage Allowance	Drafting Allowance	Pattern Total Dimension
T (thickness)	$L + U = m_1$	$(T + m_1) \times (1^*/100) = s_1$		$(T + m_1 + s_1)$
t_1 (thickness)	$U = m_2$	$(t_1 + m_2) \times (1/100) = s_2$		$(t_1 + m_2 + s_2) = p''$
t_3 (thickness)	$L = m_3$	$(t_3 + m_3) \times (1/100) = s_3$		$(t_3 + m_3 + s_3) = p'$
t_2 (thickness)	$L + U = m_1$	$(t_2 + m_1) \times (1/100) = s_4$		$(t_2 + m_1 + s_4) = p$
D_1 (upper diameter)	$2\,L_1 = m_4$	$(D_1 + m_4) \times (1/100) = s_5$	$2[p(\tan 3°)] = d_1$	$(D_1 + m_4 + s_5 + d_1)$
D_1 (lower diameter)	$2\,L_1 = m_4$	$(D_1 + m_4) \times (1/100) = s_5$		$(D_1 + m_4 + s_5)$
D_4 (upper diameter)	$2\,L_2 = m_5$	$(D_4 + m_5) \times (1/100) = s_6$	$2[p'(\tan 1.5°)] = d_2$	$(D_4 + m_5 + s_6 + d_2)$
D_4 (lower diameter)	$2\,L_2 = m_5$	$(D_4 + m_5) \times (1/100) = s_6$		$(D_4 + m_5 + s_6)$
D_2 (upper diameter)	$2\,L_2 = m_5$	$(D_2 + m_5) \times (1/100) = s_7$		$(D_2 + m_5 + s_7)$
D_2 (lower diameter)	$2\,L_2 = m_5$	$(D_2 + m_5) \times (1/100) = s_7$	$2[p''(\tan 3°)] = d_3$	$(D_2 + m_5 + s_7 + d_3)$
D_3 (hole diameter)	$2\,(-b) = m_6$	$(D_3 - m_6) \times (1/100) = s_8$		$(D_3 - m_6 + s_8)$

Notes: $L =$ allowance for surface; $U =$ allowance for cope side; $b =$ allowance for bore; $1^* =$ linear shrinkage value for gray cast iron; $m =$ machining allowance; $s =$ shrinkage allowance; $d =$ drafting allowance.

3.3 Design of Gating System—Sprue Design

A number of casting defects can be eliminated by properly designing sprue along with other gating system components (see Figure 3.1).

3.3.1 Sprue Design Rules

First, the sprue should be sized to limit the flow rate of molten metal. In small and moderately sized casting, round sprues with small height and radius should be used. An attempt should be made to locate the sprue as far from the gates as possible (see Figure 3.1); this would ensure laminar flow of molten metal. Additionally, turbulence flow can be avoided by designing the sprue so that it feeds into a standard-sized well (see Figure 3.1).

3.3.2 Mathematical Modeling of the Sprue Design

A good sprue design would ensure uniform *volumetric flow rate, Q*; this design objective is achieved by allowing a taper of minimum 5% in the sprue diameter (see Figure 3.3) so as to prevent trapped air bubbles in the molten metal stream.

It is evident from Figure 3.3 that the following mathematical relationship holds good for sprue:

$$Q = v_1 A_1 = v_2 A_2 \tag{3.1}$$

where Q is the constant volumetric flow rate, cm^3/s; v_1 is the liquid speed at the top, cm/s; A_1 is the cross-sectional area of the sprue at the top, cm^2; v_2 is the liquid speed at the bottom, cm/s; and A_2 is the cross-sectional area of the sprue at the bottom, cm^2.

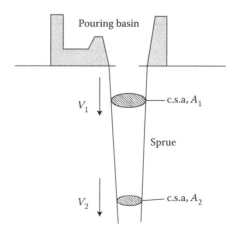

FIGURE 3.3
Taper in sprue diameter for ensuring constant Q.

Since the metal in the pouring basin is at zero velocity (reservoir assumption), the velocity of metal at the bottom of sprue can be calculated as follows:

$$v = (2gh)^{\frac{1}{2}} \tag{3.2}$$

where v is velocity of metal at sprue bottom, cm/s; h is height of sprue, cm; and $g = 981$ cm/s^2.

The time to fill a mold cavity can be calculated by

$$T_{MF} = V/Q \tag{3.3}$$

where V is the volume of mold cavity, cm^3; T_{MF} is the time to fill the mold cavity, s.

A more reliable quantitative technique for avoiding turbulent flow (preventing air entrapping defect) is the use of the *Reynolds number, R_e*, which is numerically defined as

$$R_e = (vD\rho)/\eta \tag{3.4}$$

where v is the velocity of the liquid, m/s; D is the diameter of the channel (sprue), m; ρ is the density of the liquid, kg/m^3; and η is the viscosity of the liquid at the specified temperature, Pa · s.

Equation 3.4 enables us to derive three important sprue design aspects: (1) $0 < R_e < 2000$ indicates laminar fluid flow (desirable); (2) $2000 < R_e < 20{,}000$ indicates transition flow (harmless); and (3) $R_e > 20{,}000$ indicates turbulence flow (undesirable) (air entrainment casting defects).

The significances of Equations 3.1 through 3.4 are illustrated in Examples 3.3 through 3.8.

3.4 Design of Gating System—Riser Design

Riser Design Principles: An improper riser design would result in casting defects, particularly shrinkage cavity (see Section 3.1). Hence, a proper riser design is very important in metal casting (see Figure 3.2). A good riser design must ensure that the riser cools more slowly than the casting and the riser has enough material to compensate for the casting shrinkage. According to *Chvorinov's rule,* the slowest cooling time is achieved with the greatest volume and the least surface area. This is why a cylindrical riser with a circular cross section is used in the casting design. Risers should be located near thick sections of the casting (Guleyupoglu, 2017).

Mathematical Modeling: Since a riser must cool more slowly than the casting, the following mathematical relationship has been developed by foundry engineers for a good riser design:

$$t_{\text{riser}} = 1.25 t_{\text{casting}} \tag{3.5}$$

It means that the time taken by the metal in the riser should take 1.25 times longer to freeze as compared to the casting. By taking $n = 2$ in *Chvorinov's formula*, Equation 3.5 transforms to

$$(V_{\text{riser}}/A_{\text{riser}})^2 = 1.25 \, (V_{\text{casting}}/A_{\text{casting}})^2 \tag{3.6}$$

Riser dimensions can be calculated by the following design formula (Johns, 1980; Guleyupoglu, 2017):

$$D_R{}^2 \, H_a = 24FW_c/\pi\rho \tag{3.7}$$

where D_R is the diameter of the riser, H_a is the active height of the riser, F is the feed metal requirement, W_c is the weight of the casting, and ρ is the density of the metal.

In general, a riser is slightly tapered; this design consideration can be expressed as follows:

$$V = (\pi/3) \, H_a(r^2 + rR + R^2) \tag{3.8}$$

where V is the volume of the riser, r is the small radius of the riser, and R is the large radius of the riser.

$$A = \pi(r + R) \left[H_a{}^2 + (R - r)^2 \right]^{1/2} \tag{3.9}$$

where A is the surface area of the riser, and the other symbols have their usual meanings.

3.5 Sand Mold Design

A well-designed mold would ensure elimination of many casting defects. The essential features of a good mold design include (1) the molding sand having sufficient strength, hot strength, and permeability; (2) provision of the right number of chaplets to prevent core displacement; and (3) provision of weight to prevent lifting of cope. Some important mathematical relationships in mold design are discussed in Chapter 2 (see Section 2.3.1.1). We learned in Chapter 2 that a core is subjected to gravity and buoyancy forces during the pouring of molten metal in a mold, which may cause core displacement. This technical problem is overcome by using chaplets. Chaplets are used to support a core inside a sand mold cavity. The design of a chaplet and the way the chaplet is placed in a mold cavity allows each chaplet to sustain a specified force (see Figure 2.4). Hence, the number of chaplets required to prevent core displacement can be found as follows:

$$(n_{ch})_b = W_m/F_{ch} \tag{3.10}$$

$$(n_{ch})_w = W_c/F_{ch} \tag{3.11}$$

where $(n_{ch})_b$ is the number of chaplets required above the core to resist buoyancy force; F_{ch} is the force sustained by one chaplet, N; $(n_{ch})_w$ is the number of chaplets required beneath the core to resist the weight of the core; W_m is the weight of the metal, N; and W_c is the weight of the core, N.

In Chapter 2 (Section 2.3.1.1), we considered metallostatic force (i.e., buoyancy force due to the molten metal). However, in the case of casting molds involving *cores*, there is an additional buoyancy force on the core. Hence, the net upward force on the mold can be calculated as follows:

$$F = (F_m - W_s) + F_b \tag{3.12}$$

where F is the net upward force on the upper flask (cope), N; F_m is the metallostatic force, N; W_s is the weight of the sand in the cope, N; and F_b is the net buoyancy force on the core, N. It must be noted that the weight of the core has already been considered in calculating F_b in Equation 3.12.

3.6 Mathematical Modeling in Die Casting Design

The die casting process and equipment are discussed in Chapter 2 (Section 2.3.2.3). The plunger of a die casting machine can be designed by using the following mathematical relationship (Reikher and Gerrber, 2011):

$$v = 2[(gH)^{\frac{1}{2}} - (gh)^{\frac{1}{2}}] \tag{3.13}$$

where v is the shot-sleeve velocity of the plunger (see Figure 2.7), H is the plunger diameter, h is the depth of the molten metal in the sleeve, and g is the gravitational acceleration.

3.7 Problems and Solutions—Examples in Casting Design

EXAMPLE 3.1: PATTERN DESIGN FOR METALLIC GEAR

Design the wooden pattern that is required to produce the carbon steel gear (shown in Figure E3.1) using the sand-casting process by adding the machining allowances, shrinkage allowances, and drafting allowances (all the surfaces of the part should be machined).

Solution

The pattern design requires references to Tables 3.1 and 3.2. We must select steel as the casting material and look for the machining allowance (Table 3.1) and shrinkage allowance data (Table 3.2). By reference to Figure 3.3 and the mathematical models in Table 3.3, a new Table E3.1 is now created for the pattern design.

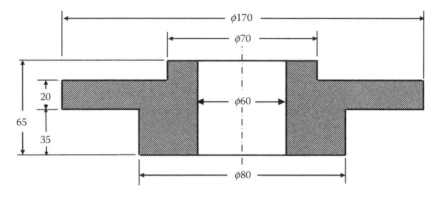

FIGURE E3.1
Dimensions of steel gear casting (all dimensions in mm).

TABLE E3.1

Pattern Allowance Calculations for the Steel Gear (All Dimensions in mm)

Nominal Dimension	Machining Allowance	Shrinkage Allowance	Drafting Allowance	Pattern Total Dimension
65 (thickness)	**3.2(L) + 6.4(U) = 9.6**	$(65 + 9.6) \times$ $(1.8/100) = 1.34$		**(65 + 9.6 + 1.34)** **= 75.9**
10 (thickness)	6.4(U) = 6.4	$(10 + 6.4) \times$ $(1.8/100) = 0.29$		$(10 + 6.4 + 0.29)$ $= 16.7$
35 (thickness)	3.2(L) = 3.2	**(35 + 3.2) ×** **(1.8/100) = 0.68**		$(35 + 3.2 + 0.68)$ $= 38.9$
20 (thickness)	**3.2(L) + 6.4(U) = 9.6**	**(20 + 9.6) ×** **(1.8/100) = 0.53**		$(20 + 9.6 + 0.53)$ $= 30.1$
170 (upper diameter)	$2 \times 4.8\ (2L_1) = 9.6$	**(170 + 9.6) ×** **(1.8/100) = 3.23**	2[30.1 (tan 3°)] = 3.13	**(170 + 9.6 + 3.23** **+ 3.13) = 186**
170 (lower diameter)	$2 \times 4.8 = 9.6$	**(170 + 9.6) ×** **(1.8/100) = 3.23**		$(170 + 9.6 + 3.23)$ $= 182.8$
80 (upper diameter)	$2 \times 3.2\ (2L_2) = 6.4$	**(80 + 6.4) ×** **(1.8/100) = 1.55**	2[38.9 (tan1.5°)] = 2.02	$(80 + 6.4 + 1.55$ $+ 2.02) = 89.9$
80 (lower diameter)	$2 \times 3.2 = 6.4$	$(80 + 6.4) \times$ $(1.8/100) = 1.55$		$(80 + 6.4 + 1.55)$ $= 87.9$
70 (upper diameter)	$2 \times 3.2\ (2L_2) = 6.4$	**(70 + 6.4) ×** **(1.8/100) = 1.37**		$(70 + 6.4 + 1.37)$ $= 77.7$
70 (lower diameter)	$2 \times 3.2\ (2L_2) = 6.4$	**(70 + 6.4) ×** **(1.8/100) = 1.37**	2[16.7 (tan 3°)] = 1.74	$(70 + 6.4 + 1.37$ $+ 1.74) = 79.5$
60 (hole diameter)	$2 \times (-3.2)\ = -6.4$	$(60 - 6.4) \times$ $(1.8/100) = 0.96$		$(60 - 6.4 + 0.96)$ $= 54.5$

EXAMPLE 3.2: PATTERN DESIGN FOR AN ALUMINUM CASTING

Design the wooden pattern that is required to produce the aluminum component (see Figure E3.2a) using the sand-casting process by calculating the changes in the nominal dimensions of the component.

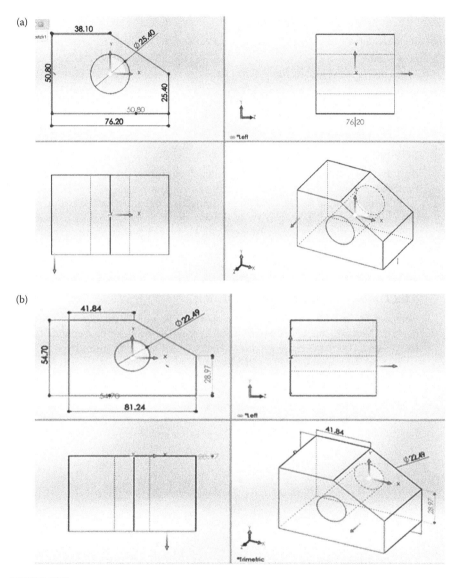

FIGURE E3.2
(a) Dimensions of aluminum casting (all dimensions in mm). (b) Dimensions of the wooden pattern (all dimensions in mm). (Adapted from Ghamdi, S.Al. 2017. *Personal communication (contribution) by Mr. Saleh Al Ghamdi, a Junior Student (during April 2017)*, Department of Mechanical Engineering, King Abdulaziz University, Jeddah, Saudi Arabia.)

TABLE E3.2

Pattern Allowance Calculations for the Aluminum Casting (All Dimensions in mm)

Nominal Dimension	Machining Allowance	Shrinkage Allowance	Pattern Total Dimension
25.4 (right-side length)	$2 \times 1.6 = 3.2$	$(25.4 + 3.2) \times (1.3/100) = 0.37$	$(25.4 + 3.2 + 0.37) = 28.97$
50.8 (left-side length)	3.2	$(50.8 + 3.2) \times (1.3/100) = 0.70$	54.7
38.1 (upper width)	3.2	$(38.1 + 3.2) \times (1.3/100) = 0.54$	41.8
76.2 (lower width)	$1.6 + 2.4 = 4$	1.04	81.2
25.4 (hole diameter)	$2 \times (-1.6) = -3.2$	$(25.4 - 3.2) \times (1.3/100) = 0.29$	22.5

Solution

By references to Tables 3.1 and 3.2, we select aluminum as the casting material. A new Table E3.2 is thus created.

By using the data in Table E3.2, an engineering drawing for the pattern can now be created. The drawing of the wooden pattern created by use of the computer-aided design software *SolidWorks* is shown in Figure E3.2b (Ghamdi, 2017).

EXAMPLE 3.3: ESTIMATING MOLTEN METAL VELOCITY AND VISCOSITY FOR DOWNSPRUE DESIGN

Refer to Figure E3.2a and suggest a reasonable height for the downsprue design. Calculate the velocity of the molten metal at the exit cross section of the downsprue. Also estimate the viscosity of molten aluminum by suggesting an appropriate pouring temperature of the metal.

Solution

By reference to the drawing of the aluminum component, the downsprue height should be in the range of 5–10 cm. We take $h = 7$ cm.

(a) By using Equation 3.2,

$$v_2 = (2gh)^{\frac{1}{2}} = (2 \times 981 \times 7)^{\frac{1}{2}} = 117.2 \, \text{cm/s}$$

The velocity of the molten metal at the exit cross section of the downsprue $= 117.2$ cm/s.

(b) The pouring temperature should be sufficiently higher than the aluminum melting temperature, $T_m = 660°C$. However, too high a pouring temperature may result in casting defects, such as *hot tears*. Hence, 750°C is an appropriate pouring temperature. By reference to the data in the literature (Chegg, 2017), the dynamic viscosity of liquid aluminum at 750°C is ~1.5 mPa · s.

EXAMPLE 3.4: COMPUTING VOLUME FLOW RATE FOR DOWN-SPRUE AND POURING BASIN DESIGNS

The gating system for a sand-casting is designed with the height of a downsprue $= 12$ cm, the diameter at the bottom $= 1.8$ cm, and the diameter at the top 2.40 cm. The diameter of the pouring cup (basin) $= 7.3$ cm. Calculate the following: (a) the velocity at the bottom of the downsprue; (b) the volume flow rate, (c) the velocity at the top of the downsprue; and (d) the vertical flow velocity within the pouring cup.

Solution

(a) By using

$$v_2 = (2gh)^{\frac{1}{2}} = (2 \times 981 \times 12)^{\frac{1}{2}} = 153.4 \, \text{cm/s}$$

The velocity at the bottom of the downsprue $= 153.4$ cm/s.

(b) Cross-sectional area of downsprue at the bottom $= A_2 = (\pi/4) \, d_2^2 = (\pi/4) \, (1.8)^2 = 2.54 \, \text{cm}^2$.

By using Equation 3.1,

$$Q = v_2 A_2 = (153.4 \, \text{cm/s}) \, (2.54 \, \text{cm}^2) = 390.3 \, \text{cm}^3/\text{s}$$

The volume flow rate $= 390.3 \, \text{cm}^3/\text{s}$.

(c) Cross-sectional area of downsprue at the top $= A_1 = (\pi/4) \, d_1^2 = (\pi/4) \, (2.4)^2 = 4.52 \, \text{cm}^2$.

$$Q = v_1 A_1 \rightarrow v_1 = (Q/A)_1 = (390.3 \, \text{cm}^3/\text{s})/(4.52 \, \text{cm}^2) = 86.3 \, \text{cm/s}$$

The velocity at the top of the downsprue $= 86.3$ cm/s.

(d) Cross section of the pouring cup $= A = (\pi/4) \, d^2 = (\pi/4) \, (7.3)^2 = 41.85 \, \text{cm}^2$.

$$Q = vA \rightarrow v = Q/A = (390.3)/(41.85) = 9.32 \, \text{cm/s}$$

The vertical flow velocity within the pouring cup $= 9.32$ cm/s.

EXAMPLE 3.5: DESIGNING A DOWNSPRUE BY SPECIFYING ITS DIAMETER

The flow rate of liquid metal into the downsprue of a sand mold is $800 \, \text{cm}^3/\text{s}$. The downsprue is 160 mm long. Calculate the base diameter of the sprue to maintain the uniform flow rate of the metal.

Solution

$Q = 800 \, \text{cm}^3/\text{s}; h = 160 \, \text{mm} = 16 \, \text{cm}$.

By using Equation 3.2,

$$v_2 = (2gh)^{\frac{1}{2}} = (2 \times 981 \times 16)^{\frac{1}{2}} = 177.1 \, \text{cm/s}$$

By rearranging the terms in Equation 3.1,

$$A_2 = Q/v_2 = (800)/(177.1) = 4.5\,\text{cm}^2$$

$$A_2 = (\pi/4)\,D^2; \quad 4.5 = (\pi/4)\,D^2$$

$$D^2 = (4 \times 4.5)/\pi = 5.75\,\text{cm}^2; \quad D = 2.4\,\text{cm}$$

Diameter at the base of the downsprue $= 2.4$ cm.

EXAMPLE 3.6: COMPUTING TIME TO FILL A MOLD CAVITY FOR DOWNSPRUE DESIGN

The downsprue of a casting gating system has a length of 18 cm. The cross-sectional area at the bottom of the sprue is 4 cm^2. The mold cavity has a volume 1050 cm^3. Compute the following: (a) the velocity of the molten metal at the bottom of the downsprue, (b) volume flow rate, and (c) the time required to fill the mold cavity. Is the computed time value desirable by a foundry engineer?

Solution

(a) By using Equation 3.2,

$$v_2 = (2gh)^{\frac{1}{2}} = (2 \times 981 \times 18)^{\frac{1}{2}} = 187.9\,\text{cm/s}$$

The velocity of the molten metal at the bottom of the downsprue $=$ 187.9 cm/s.

(b) By using Equation 3.1,

$$Q = v_2\,A_2 = (187.9) \times (4) = 751.7\,\text{cm}^3/\text{s}$$

The volume flow rate $= 751.7$ cm^3/s.

(c) By using Equation 3.3,

$$T = V/Q = (1050\,\text{cm}^3)/(751.7\,\text{cm}^3/\text{s}) = 1.4\,\text{s}$$

The time to fill the mold cavity $= 1.4$ s.

The computed time value is desirable because the time to fill the mold cavity is reasonably short.

EXAMPLE 3.7: DESIGNING A SPRUE BY COMPUTING THE REYNOLDS NUMBER

A sand-casting gating system is to be designed with the height of down-sprue $= 13$ cm, and sprue diameter $= 0.7$ cm. The density of molten metal is 3.4 g/cm^3, and its viscosity is 0.001 Pa · s. Compute the Reynolds number. Is the sprue design appropriate for a defect-free sand-casting? If not, what measures should be taken to improve the sprue design (without changing the data)?

Solution

$h = 13$ cm $= 0.13$ m, $D = 0.7$ cm $= 7 \times 10^{-3}$ m, $\rho = 3.4$ g/cm^3 $= 3400$ kg/m^3, $\eta = 0.001$ Pa $\cdot$ s.

By using Equation 3.2,

$$v_2 = (2gh)^{\frac{1}{2}} = (2 \times 9.81 \times 0.13)^{\frac{1}{2}} = 1.6 \, \text{m/s}$$

The velocity of the molten metal in the channel (v) will be the same as the velocity at the bottom of the downsprue, v_2.

By using Equation 3.4,

$$R_e = (vD\rho)/\eta = [(1.6) \, (7 \times 10^{-3}) \, (3400)]/(10^{-3}) = 38{,}010$$

Since the Reynolds number $R_e > 20{,}000$, the fluid flow is *turbulent* (resulting in casting defects). Hence, the sprue design is bad. In order to improve the sprue design, a well at the base of the downsprue should be provided (see Figure 3.2); this design improvement would ensure laminar fluid flow rather than turbulent flow.

EXAMPLE 3.8: DESIGNING A SPRUE FOR AN ALUMINUM CASTING

Design the sprue for the aluminum casting shown in Figure E3.2a.

Solution

In order to create a good design for the sprue for the casting of the aluminum component shown in Figure E3.2a, we should ensure that the Reynolds number < 2000. Assume $R_e = 1900$.

Now we refer to the data in Example 3.3:

$$v = 117.2 \, \text{cm/s} = 1.17 \, \text{m}, \, \eta_{Al} = 1.5 \times 10^{-3} \, \text{Pa} \cdot \text{s}, \, \rho_{Al} \text{ at } 750°\text{C}$$
$$= 2350 \, \text{kg/m}^3 \, (\text{Chegg, 2017})$$

By using Equation 3.4,

$$R_e = (vD\rho)/\eta$$

$$1900 = [(1.17) \, (D) \, (2350)]/(1.5 \times 10^{-3})$$

$$D = 1.03 \times 10^{-3} \, \text{m} \sim 1 \, \text{mm}$$

The computed sprue diameter is too small to be practical. This impracticality in the sprue design calls us to increase sprue (channel) diameter to at least 5 mm (= 0.005 m). Now, we iterate with $D = 0.005$ m and calculate R_e.

By using Equation 3.4,

$$R_e = (vD\rho)/\eta = [(1.17)\,(0.005)\,(2350)]/(1.5 \times 10^{-3})$$
$$= 9.160 \times 10^3 = 9160$$

Since the Reynolds number $2000 < R_e < 20{,}000$, the fluid flow is *laminar* (desirable). Hence, the new design (with $D = 5\,\text{mm}$) is the optimum sprue design.

EXAMPLE 3.9: CONFIRMING A RISER DESIGN WHEN (V/A) RATIOS ARE GIVEN

The (V/A) ratio of a casting is 7.8. A riser designed for this casting has $(V/A) = 9$. Is the riser design good?

Solution

$$(V_{\text{casting}}/A_{\text{casting}})^2 = (7.8)^2 = 60.8\,\text{cm}^2$$

$$(V_{\text{riser}}/A_{\text{riser}})^2 = 9^2 = 81$$

$$[(V_{\text{riser}}/A_{\text{riser}})^2]/[(V_{\text{casting}}/A_{\text{casting}})^2] = 1.33$$

$$(V_{\text{riser}}/A_{\text{riser}})^2 = 1.33\,[(V_{\text{casting}}/A_{\text{casting}})^2]$$

By reference to Equation 3.6, we conclude that the *riser design is good*.

EXAMPLE 3.10: DESIGNING A RISER WHEN THE CASTING DIMENSIONS AND SOLIDIFICATION TIME ARE GIVEN

The solidification time for a steel casting having dimensions (14 cm × 8 cm × 2 cm) is 1.4 min. A cylindrical riser with a diameter-to-height ratio of 1 is to be designed for a sand-casting gating system. Using $n = 2$, design the riser.

Solution

First, the steel casting ($t_s = 1.4\,\text{min}$) is considered:

$$V_c = 14\,\text{cm} \times 8\,\text{cm} \times 2\,\text{cm} = 224\,\text{cm}^3$$

$$A_c = 2\,[(14 \times 8) + (8 \times 2) + (2 \times 14)] = 2\,(112 + 16 + 28) = 312\,\text{cm}^2$$

By using Equation 2.7,

Mold constant for the steel casting $= B = t_s/(V/A)^n = 1.4\,/(224/312)^2 = 1.4\,/\,0.51 = 2.71\,\text{min/cm}^2$.

Now, the riser is designed by specifying its dimensions as follows:

$$\text{Volume of riser} = V_r = (\pi d^2 h)/4$$
$$\text{Surface area of riser} = A_r = \pi dh + (\pi d^2/2)$$

Since the ratio $(d/h) = 1$, $d = h$, the volume and surface area expressions become

$$V_r = \pi d^3/4 \quad \text{and} \quad A_r = 3\pi d^2/2$$

The ratio simplifies to $V_r/A_r = d/6$.

By using Equation 3.5,

$$t_{\text{riser}} = 1.25 t_{\text{casting}} = 1.25 \times 1.4 = 1.75 \min$$

Again by using Equation 2.7 for the riser,

$$t_s = B\,(V/A)^2, \ 1.75\min = (2.71\min/\text{cm}^2)\,(d/6)^2$$

$$d = 4.8\,\text{cm}$$

Hence, the riser design should have diameter of the riser $=$ height $=$ 4.8 cm.

EXAMPLE 3.11: CONFIRMING DESIGN OF A TAPERED RISER

A metal casting having volume of $32 \times 10^6 \text{ mm}^3$ and surface area of $76 \times 10^4 \text{ mm}^2$ is to be manufactured by sand-casting process. The riser design involves the following dimensions:

Riser small diameter $= 150$ mm; Riser large diameter $= 300$ mm; Riser height $= 200$ mm.

Is the riser design correct for the casting?

Solution

$$(V_{\text{casting}}/A_{\text{casting}}) = (32{,}000{,}000\text{ mm}^3)/(760{,}000\text{ mm}^2) = 42.1\text{ mm}$$

$$(V_{\text{casting}}/A_{\text{casting}})^2 = 1772.8\text{ mm}^2$$

Riser

$r = 75$ mm; $R = 150$ mm; $h = 200$ mm.

By using Equation 3.8,

$$V = (\pi/3)\,H_a\,(r^2 + rR + R^2) = (\pi/3)\,(200)\,[(75)^2 + (75)\,(150) + (150)^2]$$
$$= 8{,}245{,}125\text{ mm}^3$$

By using Equation 3.9,

$$A = \pi\,(r + R)\,[H_a{}^2 + (R - r)^2]^{\frac{1}{2}} = \pi(75 + 150)\,[(200)^2 + (150 - 75)^2]^{\frac{1}{2}}$$

$$= 150{,}985\text{ mm}^2$$

$$(V/A) = 54\text{ mm (approx)} \quad \text{or} \quad (V/A)^2 = 2982\text{ mm}^2$$

$$[(V_{\text{riser}}/A_{\text{riser}})^2]/[(V_{\text{casting}}/A_{\text{casting}})^2] = (2982)/(1764) = 1.69$$

$$(V_{\text{riser}}/A_{\text{riser}})^2 = 1.69\,(V_{\text{casting}}/A_{\text{casting}})^2$$

By referring to Equation 3.6, we should have $(V_{riser}/A_{riser})^2 = 1.25$ $(V_{casting}/A_{casting})^2$.

The riser design is not perfect; however, it is acceptable.

EXAMPLE 3.12: SAND MOLD DESIGN—COMPUTING NUMBER OF CHAPLETS

A sand mold is to be designed for producing a copper casting with a bore. The design of the chaplets and the manner in which they are placed in the mold cavity permit each chaplet to sustain a force of 47 N. Several chaplets are located beneath the core; several other chaplets are placed above the core. The core has a volume of 5080 cm³. Determine the number of chaplets that should be placed (a) beneath the core and (b) above the core.

Solution

By reference to Table 2.1, $\rho_{copper} = 8.9$ g/cm³ $= 8900$ kg/m³; $\rho_{sand} = 1.6$ g/cm³ $= 1600$ kg/m³.

Volume of core = Volume of metal displaced by the core $= V = 5080$ cm³ $= 0.00508$ m³.

Force sustained by a chaplet $= F_{ch} = 47$ N.

(a) Let weight of the metal displaced by the core during poring $= W_m$:

$$W_m = (\rho_{copper})\, Vg = 8900 \times 0.00508 \times 9.81 = 443.5\,\text{N}$$

By using Equation 3.10,

$$(n_{ch})_b = W_m/F_{ch} = (443.5)/(47) = 9.4 \sim 10$$

Hence, *10 chaplets* are required *above the core* to resist buoyancy force.

(b) Weight of core $= W_c = (\rho_{sand})\, Vg = 1600 \times 0.00508 \times 9.81 = 79.7$ N.

By using Equation 3.11,

$$(n_{ch})_w = W_c/F_{ch} = (79.7)/(47) = 1.69 \sim 2$$

Hence, *two chaplets* are required *beneath the core* to resist the weight of the core.

EXAMPLE 3.13: MOLD DESIGN—COMPUTING WEIGHT REQUIRED TO PREVENT THE FLASK (COPE) FROM LIFTING

A rectangular cast iron component having dimensions (21 cm × 21 cm × 20 cm) with a bore is required to be manufactured by sand-casting. A sand core located inside a mold cavity has a volume of 2450 cm³. If the flask dimensions are (320 × 320 × 200 mm³), calculate (a) the metallostatic force, and the weight of the sand in the cope; (b) the net buoyancy force on the core; and (c) the weight required to prevent the flask (cope) from lifting up.

Solution

(a) By reference to Example 2.5,

The metallostatic force $= F_m = 620$ N.

The weight of the sand in the cope $= W_s = 335.5$ N.

(b) By reference to Example 2.6,

The buoyancy force on the core $= F_b = 133.6$ N.

(c) By using Equation 3.12,

$$F = (F_m - W_s) + F_b = (620 - 335.5) + 133.6 = 418.10 \, \text{N} \sim 420 \, \text{N}$$

The net upward force $= 420$ N.

$$W = mg, \, 420 = m \, (9.81)$$
$$m = 42.8 \, \text{kgf}$$

Hence, the weight required to prevent the flask (cope) from lifting up $= 43$ kgf.

EXAMPLE 3.14: DIE CASTING MACHINE'S PLUNGER DESIGN

The slow shot velocity of the plunger for a cold-chamber die casting machine is designed to be 0.3 m/s. The die casting process requires the depth of the molten metal in the shot sleeve to be 6 cm. Design the plunger by specifying its diameter.

Solution

$h = 6 \, \text{cm} = 0.06 \, \text{m}; \, v = 0.3 \, \text{m/s}; \, g = 9.81 \, \text{m/s}^2.$

By using Equation 3.13,

$$v = 2[(gH)^{\frac{1}{2}} - (gh)^{\frac{1}{2}}]$$
$$0.3 = 2\,[(9.81H)^{\frac{1}{2}} - (9.81 \times 0.06)^{\frac{1}{2}}]$$
$$H = 0.09 \, \text{m}$$

Plunger diameter $= 9$ cm.

Questions and Problems

3.1 List the factors that may cause casting defects.

3.2 You, as a foundry engineer, found a strong crust of fused sand on the surface of a casting. What measures should you take to avoid this casting defect?

3.3 Besides pattern allowances, what are the other considerations in pattern design?

P3.4 Design the wooden pattern that is required to produce the cast iron gear (shown in Figure E3.1) using the sand-casting process by adding the machining allowances, shrinkage allowances, and drafting allowances (all the surfaces of the part should be machined).

3.5 Why is a cylindrical riser with a circular cross section used in a gating system of casting design?

3.6 Sketch a sand mold showing all components of the gating system and the mold features for producing a casting with a bore.

P3.7 The flow rate of liquid metal into the downsprue of a sand mold is 700 cm^3/s. The downsprue is 140 mm long. What diameter of the sprue should be used at its base to maintain the uniform flow rate of the molten metal?

P3.8 A sand-casting gating system is to be designed by using a 15 cm long downsprue having sprue diameter $= 0.5$ cm. The density of molten metal is 5.5 g/cm^3, and its viscosity is 2 mPa · s. Compute the Reynolds number. Is the sprue design appropriate for a defect-free sand-casting? If not, what measures should be taken to improve the sprue design (without changing the data)?

P3.9 The downsprue of a casting gating system has a height of 13 cm. The cross-sectional area at the bottom of the sprue is 2.8 cm^2. The mold cavity has a volume 950 cm^3. Compute (a) the velocity of the molten metal at the bottom of the downsprue, (b) the volume flow rate, and (c) the time required to fill the mold cavity.

P3.10 The solidification time for a steel casting having dimensions (16 cm $\times$ 9 cm $\times$ 3 cm) is 1.8 min. A cylindrical riser with a diameter-to-height ratio of 1 is to be designed for a sand-casting gating system. Using $n = 2$, design the riser.

P3.11 A cold-chamber die casting machine's plunger has a diameter of 9 cm. The slow shot velocity of the plunger is 0.34 m/s. Calculate the depth of the molten metal in the shot sleeve.

References

Campbell, J. 2011. *Complete Casting Handbook: Metal Casting Processes, Techniques and Design.* Oxford, UK: Butterworth-Heinemann/Elsevier Science.

Chegg. 2017. Internet Source: http://www.chegg.com/homework-help/questions-and-answers/molten-pure-aluminum-750-c-flows-flat-plate-silicon-also-750-c–free-stream-velocity-02-m–q12819951.

Ghamdi, S.Al. 2017. *Personal communication (contribution) by Mr. Saleh Al Ghamdi, a Junior Student (during April 2017)*, Department of Mechanical Engineering, King Abdulaziz University, Jeddah, Saudi Arabia.

Goodrich, G. 2008. *Casting Defects Handbook: Iron and Steel*. Schaumburg, IL: American Foundry Society.

Groover, M.P. 2014. *Fundamentals of Modern Manufacturing*. New York: John Wiley and Sons.

Guleyupoglu, S. 2017. Casting Process Design Guidelines; Internet Source: http://www.metalwebnews.com/image/casting-guidelines.pdf.

Johns, R.A. 1980. Risering steel castings easily and efficiently. *AFS Transactions*, 88, 77–84.

Reikher, A. Gerrber, H. 2011. Calculation of the Die Cast Parameters of the Thin Wall Aluminum Die Cast Part. 2011 Die Casting Congress and TableTop. Wheeling, IL: NADCA (Internet Source: https://www.flow3d.com/wp-content/uploads/2014/08/Calculation-of-the-Die-Cast-parameters-of-the-Thin-Wall-Aluminum-Cast-Part.pdf).

4

Plastic Molding Processes

4.1 Viscoelasticity and Die Swell

General: Viscoelasticity refers to a material property of viscous-like as well as elastic characteristic of materials. Viscoelasticity is exhibited by almost all polymers, especially by liquid polymers.

The extrusion of plastic parts involves a great deal of viscoelasticity that is manifested by *die swell* (Koopmans, 1999). The expansion of a hot plastic part when it exits the die opening is called *die swell* (see Figure 4.1). The die swell problem can be overcome by using a longer channel in the extrusion machine.

Mathematical Modeling: In circular cross-sectional polymeric parts, die swell can be measured by the *swell ratio*, which is defined as the ratio of the diameter of the extruded cross section to the diameter of the die orifice (see Figure 4.1). Mathematically, swell ratio, r_s, is given by

$$r_s = D_x/D_d \tag{4.1}$$

where D_x is the diameter of the extruded cross section, mm; and D_d is the diameter of the die orifice, mm.

4.2 Extrusion Molding of Plastics

Process: The extrusion molding process involves extruding plastic by forcing a continuous stream of molten material through an open die (see Figure 4.2). The diameter of the die determines the diameter of the product. In extrusion molding, thermoplastic molding power is fed from a hopper. Then the (helical) power screw forces the plastic material that is continually thread through an open die to form a continuous part with constant cross-sectional shape (Figure 4.2). The heat and friction cause plasticizing and softening of the material and forces it through the die opening. Finally, the material is cooled by either air or water, and the part is removed (Rauwendaal, 2014). Examples of extruded parts include polyvinyl chloride (PVC) pipes, sheets, and the like.

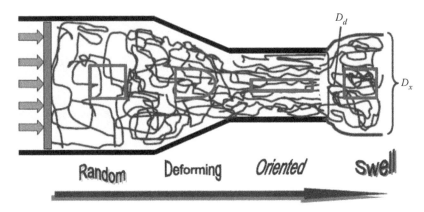

FIGURE 4.1
Die swell in plastic (extrusion) molding.

Mathematical Modeling: Since the extruder screw (inside the barrel) is the driving element in the extrusion machine, it is important to first mathematically model the extruder screw (see Figure 4.3). The operation of the helix screw is determined by its design parameters and the speed of rotation.

It is evident in Figure 4.3 that the pitch (p), channel depth (d_c), helix angle (ϕ), screw diameter (D), and flight width are the most important parameters in the screw geometric design. There is a strong dependence of the screw flight pitch on the screw helix angle, which is given by

$$\tan \phi = p/(\pi D) \tag{4.2}$$

where ϕ is the helix angle of the screw, p is the screw flight pitch, and D is the screw diameter. The significance of Equation 4.2 is illustrated in Example 4.1.

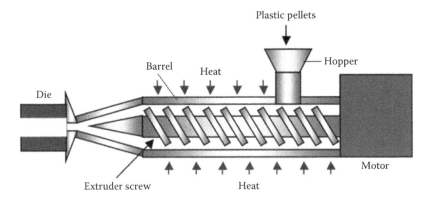

FIGURE 4.2
Extrusion molding of plastic.

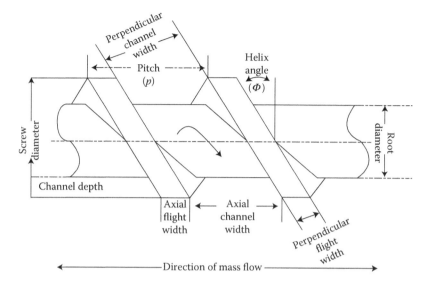

FIGURE 4.3
The geometric design parameters of an extruder screw.

Now the fluid low in the barrel is considered. Since there exists friction between the viscous liquid and two opposing surfaces moving relative to each other, the fluid flow in the barrel involves *drag flow*. Figure 4.4 illustrates that the velocity of the fluid increases from 0 to its maximum velocity, v. Thus, the average velocity of the fluid is $(0 + v)/2 = 0.5\ v$.

An important process parameter in extrusion molding is the *drag flow rate* of the semisolid plastic. The determination of *drag flow rate* requires mathematically modeling the flow of polymeric semisolid inside the barrel.

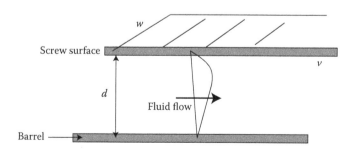

FIGURE 4.4
Drag flow of fluid in the barrel. (Note: the barrel surface is stationary while the screw surface is moving.)

By reference to Figure 4.4, the volume drag flow rate of the fluid in the barrel can be expressed as

$$Q_d = v_{av}dw = 0.5vdw \tag{4.3}$$

where Q_d is the volume drag flow rate of the fluid (semisolid plastic), m^3/s; v is the velocity of the screw surface, m/s; d is spacing between the barrel and screw surfaces, m; and w is the width of the screw (or barrel) measured perpendicular to the velocity direction, m.

The parameters in Equations 4.2 and 4.3 may be linked to those in the channel defined by the rotating screw and the stationary barrel surface (see Figures 4.3 and 4.4); this link leads to

$$d = d_c \tag{4.4}$$

$$v = \pi DN \cos \phi \tag{4.5}$$

and

$$w = w_c = (\pi D \tan \phi - w_f)\cos \phi \tag{4.6}$$

where w is the screw flight width, m; w_c is the screw channel width, m; w_f is the flight land width; D is the screw flight diameter, m; N is the screw rotational speed, rev/s; ϕ is the screw helix angle; and d_c is the screw channel depth, m.

Since the flight land width, w_f, is too small, it may be ignored; thus, Equation 4.6 simplifies to

$$w = \pi D(\tan \phi)(\cos \phi) = \pi D(\sin \phi) \tag{4.7}$$

By substituting the values of d, v, and w from Equations 4.4, 4.5, and 4.7 into Equation 4.3, we get

$$Q_d = 0.5\pi^2 D^2 N d_c (\sin \phi)(\cos \phi) \tag{4.8}$$

Hence, the *volume drag flow rate* of the fluid, Q_d, can be computed when the other design and operating parameters are given.

4.3 Blow Molding of Plastics

Process: Blow molding involves the manufacture of hollow plastic parts by inflating a heated plastic tube until it fills a mold cavity and forms the desired shape. The raw material is a thermoplastic in the form of small granules, which is first melted and formed into a hollow tube, called the *parison*

(Figure 4.5[a]). The parison is then clamped between two mold halves and inflated by pressurized air until it conforms to the inner shape of the mold cavity (Figure 4.5[b]). Once the part has cooled, the mold halves are separated, and the part is ejected (Figure 4.5[c]). Blow molded products include plastic hollow and thin-walled bottles and containers.

Mathematical Modeling: The mathematical modeling of *extrusion blow molding* begins with the application of the concept of die swell (see Section 4.1). The mean diameter of the parison tube, as it exits the die, may be represented by the mean diameter D_d (see Figures 4.1 and 4.5[a]). The die swell phenomenon causes expansion of parison resulting in an increase of mean diameter from D_d to D_x; this is accompanied by an increase of parison wall thickness from t_d to t_x. Since the diameter of swelling is proportional to the thickness of swelling, Equation 4.1 can be rewritten as

$$r_s = D_x/D_d = t_x/t_d \tag{4.9}$$

where t_d is the wall thickness of the parison tube as it exits the extruder die, mm; and t_x is the wall thickness of the extruded parison, mm.

When the parison is inflated, the diameter of the molding increases from D_x to D_m (see Figure 4.5[b]). Since the volume of the molding remains constant, the wall thickness of the molding decreases from t_x to t_m. Hence, the *blow in* stage can be expressed as

$$\text{Volume}_{\text{pre-blow}} = \text{Volume}_{\text{post-blow}}$$

$$\pi D_x t_x = \pi D_m t_m \tag{4.10}$$

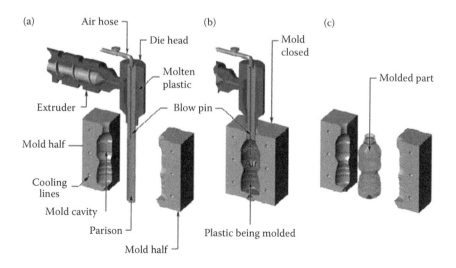

FIGURE 4.5
Extrusion *blow molding*: (a) parison extrusion, (b) blow in, and (c) molded part removal.

where D_x is the pre-blow diameter of the molding; D_m is the post-blow diameter of the molding; t_x is the pre-blow wall thickness; and t_m is the post-blow wall thickness of the molding.

By solving Equation 4.10 for t_m, we get

$$t_m = (D_x t_x)/D_m \tag{4.11}$$

By substituting the value of t_x from Equation 4.9 into Equation 4.11, we obtain

$$t_m = (r_s^2 t_d D_d)/D_m \tag{4.12}$$

where the terms have their usual meanings.

Blown-Film Extrusion Process and Its Mathematical Modeling: Polyethylene (plastic shopping) bags are manufactured by the blown-film extrusion process. In this process, small plastic pellets are melted down under controlled conditions. The molten mass is then extruded through a circular die gap to form a continuous tube of plastic. The tube (in molten state) is pinched off at one end and then inflated (expanded) from 1.5 to 2.5 times the extrusion-die diameter (Kalpakjian and Schmid, 2013). Finally, the expanded tube is stretched to the size and thickness of the desired finished product. The extrusion-die diameter can be computed by the relationship

$$d_d = D/2.5 \tag{4.13}$$

where d_d is the extrusion-die diameter, mm; and D is the blown diameter, mm. The blown diameter can be related to the perimeter of the expanded tube, as follows:

$$\text{Perimeter} = \pi D \tag{4.14}$$

If the lateral dimension (width) of a plastic shopping bag (made by blown-film extrusion process) is w, the perimeter of the flat bag can be expressed as (Kalpakjian and Schmid, 2013)

$$\text{Perimeter} = 2w \tag{4.15}$$

By combining Equations 4.13 through 4.15, we obtain

$$d_d = (2w/\pi)/2.5 = (2w)/(2.5\pi) \tag{4.16}$$

where d_d is the extrusion-die diameter, mm; and w is the lateral dimension (width) of the plastic shopping bag, mm (see Example 4.11).

4.4 Injection Molding of Plastics

4.4.1 Injection Molding Process

In injection molding, either a thermoplastic or thermosetting plastic is heated to a highly plastic state and then injected to a mold cavity by using a feed screw similar to one used in extrusion molding. The mold cavity, into which molten plastic is injected at high pressure, is the inverse of the desired shape of the part to be molded (Kazmer, 2007). The mold is held closed under high pressure and cooled so as to allow the rapid solidification of the molding (Mitani, 1997) (see Figure 4.6). Once the plastic molding has cooled, the mold is opened, usually automatically, and the finished product is removed. Examples of injection molded plastic products include bottle caps, combs, wire spools, automotive dashboards, and the like.

Significant shrinkage may occur during cooling of the plastic in the mold. Shrinkage of a given polymer is usually measured as the reduction in linear size during cooling from the molding temperature to room temperature. The typical shrinkage values for some polymers, expressed in millimeters/ millimeters, are given in Table 4.1.

4.4.2 Mathematical Modeling of Injection Molding

Shrinkage Effect: In view of the shrinkage effect of the plastic molding as discussed in the preceding paragraph, it is important that the dimensions of

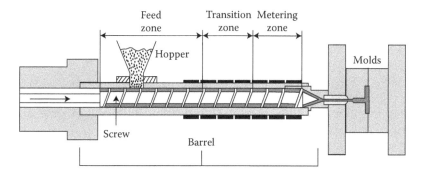

FIGURE 4.6
Injection molding of plastic.

TABLE 4.1

Typical Shrinkage Values for Some Polymers

Plastic Type	PVC	Polyethylene	Nylon 6-6	ABS	Polycarbonate
Shrinkage, mm/mm	0.005	0.025	0.020	0.006	0.007

the mold cavity must be greater as compared to the dimensions of the product. The dimensions of the mold cavity can be determined by

$$D_c = D_m + (D_m S) + (D_m S^2) \tag{4.17}$$

where D_c is the dimension of the cavity, mm; D_m is the dimension of the molded part, mm; and S is the shrinkage value obtained from the data in Table 4.1.

Cooling Time: The minimum cooling time required until the average temperature of the injection molded product reaches the mold wall temperature (usually 50°C) is calculated by using the following mathematical model (Kimerling, 2002):

$$t_c = (h^2/\pi^2\alpha)\{\ln[(4T_M - 4T_W)/(\pi T_E - \pi T_W)]\} \tag{4.18}$$

where t_c is the cooling time, s; h is the wall thickness of the molded product, mm; α is the thermal diffusivity, mm^2/s; T_M is the melt temperature, °C; T_W is the mold wall temperature, °C; and T_E is the ejection temperature, °C.

Clamping Force: The design of a new mold for an injection molding machine demands the calculation of the optimum required mold clamping force that the injection molding machine must have for the mold to be installed in it. This clamping force can be computed by the formula

$$F = (pA)/1000 \tag{4.19}$$

where F is the required mold clamping force, tf; p is the pressure inside the cavity, kgf/cm^2; and A is the total projection area, cm^2.

Three-Zone Screw—Feed Zone Calculation: The extruder screw of a conventional plasticating extruder (either in an extrusion molding or in an injection molding machine) has three zones: (1) feed zone, (2) transition zone, and (3) metering zone (see Figure 4.6).

The solids conveying rate in the feed zone of the extruder screw can be calculated according to the formula (Rao and Schott, 2012)

$$G = 60\rho N\eta_F\pi^2 HD(D - H)[W/(W + w_{FLT})](\sin\phi)\,(\cos\phi) \tag{4.20}$$

where G is the solids conveying rate, kg/h; ρ is the bulk density of the polymer, kg/m^3; N is the screw rotational speed, rpm; η_F is the conveying efficiency; H is the depth of the feed zone, m; D is the screw diameter, m; W is the channel width, m; w_{FLT} is the flight width, m; and ϕ is the screw helix angle (see Figure 4.3 and Example 4.12).

Injection Speed: The machine pump is usually expressed as oil volume flow rate (gal./min). The *injection speed*, or the *piston-travel speed*, can be calculated by the formula

$$v_p = Q_o/(60A_p) \tag{4.21}$$

where v_p is the piston speed, in/s; Q_o is the oil volume flow rate, in^3/min; and A_p is the piston cross-sectional area, in^2 (see Example 4.13).

4.5 Rotational Molding of Plastics

Rotational molding is a process for manufacturing mostly hollow plastic items. In *rotational molding*, a powder or liquid resin is placed into a hollow mold, which is then rotated bi-axially in an oven (Isayev, 2016).

4.6 Compression Molding

In the *compression molding* process, a measured amount of thermoset powder, granules, is fed into the mold cavity, where the plastic is heated under pressure. Heat and pressure allow the material to soften and flow to fill the mold cavity. Finally, the mold is opened, and the part is removed.

4.7 Problems and Solutions—Examples in Plastic Molding

EXAMPLE 4.1: DETERMINING THE HELIX ANGLE OF A SQUARE PITCH EXTRUDER SCREW

A square pitch screw has a pitch=screw diameter. An extrusion machine has a square pitch extruder screw. Compute the helix angle of the screw.

Solution

$p = D$, helix angle $= \phi = ?$

By using Equation 4.2,

$$\tan \phi = p/(\pi D) = 1/\pi$$

$$\phi = \tan^{-1}(1/\pi) = 17.6°$$

Hence, the helix angle of the extruder screw $= 17.6°$.

EXAMPLE 4.2: COMPUTING THE VELOCITY OF THE SCREW SURFACE IN AN EXTRUDER BARREL

An extruder barrel has a diameter of 8 cm. The screw rotates at 1.2 rev/s. The screw flight angle is 18°. Compute the velocity of the screw surface in the stationary barrel.

Solution

$D = 8\,\text{cm} = 0.08\,\text{m}$, $N = 1.2\,\text{rev/s}$, $\phi = 18°$

By using Equation 4.5,

$$v = \pi D N \cos \phi = \pi(0.08)(1.2)\cos 18° = 0.28\,\text{m/s}$$

Hence, the velocity of the screw surface $= 28\,\text{cm/s}$.

EXAMPLE 4.3: CALCULATING THE SCREW FLIGHT WIDTH OF EXTRUDER SCREW

Using the data in Example 4.2, calculate the screw flight width of the extruder screw.

Solution

$D = 8\,\text{cm}$, $\phi = 18°$

By using Equation 4.7,

$$w = \pi D \sin \phi = \pi(8)(\sin 18°) = 7.76\,\text{cm}$$

The screw flight width of the extruder screw $= 7.76\,\text{cm}$.

EXAMPLE 4.4: DETERMINING VOLUME DRAG FLOW RATE OF PLASTIC IN EXTRUSION MOLDING

An extrusion molding machine's barrel has a diameter of 8.5 cm. The screw rotates at 1.3 rev/s. The channel depth is 7.0 mm, and the screw flight angle is 20°. Calculate the volume drag flow rate of the plastic in the barrel of the extrusion machine.

Solution

$D = 8.5\,\text{cm} = 8.5 \times 10^{-2}\,\text{m}$, $N = 1.3\,\text{rev/s}$, $d_c = 7.0\,\text{mm} = 7.0 \times 10^{-3}\,\text{m}$, $\phi = 20°$, $Q_d = ?$

By using Equation 4.8,

$$Q_d = 0.5\pi^2 D^2 N d_c \sin \phi \cos \phi = 0.5\pi^2 (8.5 \times 10^{-2})^2 (1.3)(7.0 \times 10^{-3})$$
$$\times \sin 20° \cos 20°$$

$$= (0.5)(9.87)(72.25)(10^{-7})(9.1)(0.342)(0.94) = 1042.74 \times 10^{-7}$$
$$= 10.43 \times 10^{-5}$$

The volume drag flow rate $= 10.43 \times 10^{-5}\,\text{m}^3/\text{s}$.

EXAMPLE 4.5: COMPUTING THE SWELL RATIO FOR AN EXTRUDER DIE

The extrusion die, for plastic parison used in blow molding, has a mean diameter of 17 mm. The diameter of the ring opening in the extrusion die is 2 mm. The die swell results in an increase of the mean diameter of the parison to 20 mm after exiting the die orifice. Compute the swell ratio.

Solution

By reference to the die swell phenomenon discussed in Section 4.3, we can write

$$D_d + t_d = 17 \text{ mm} \tag{E4.5.1}$$

$$D_d - t_d = 2 \text{ mm} \tag{E4.5.2}$$

solving Equations E4.5.1 and E4.5.2 simultaneously, we obtain

$D_d = 9.5$ mm, and $t_d = 7.5$ mm; $D_x = 20$ mm

By using Equation 4.1,

$$\text{Swell ratio} = r_s = D_x/D_d = (20)/(9.5) = 2.1$$

EXAMPLE 4.6: COMPUTING THE WALL THICKNESS OF EXTRUDED PARISON FOR BLOW MOLDING

Using the data in Example 4.5, calculate the wall thickness of the extruded parison.

Solution

$t_d = 7.5$ mm, $r_s = 2.1$, $t_x = ?$

By using Equation 4.9,

$$t_x = r_s \cdot t_d = 2.1 \times 7.5 = 15.7 \text{ mm}$$

The wall thickness of the extruded *parison* = 15.7 mm.

EXAMPLE 4.7: COMPUTING THE WALL THICKNESS OF A PLASTIC BLOW MOLDING

The extrusion die, for plastic parison used in blow molding, has a mean diameter of 17.0 mm. The diameter of the ring opening in the die is 2.0 mm. The die swell results in an increase of the mean diameter of the parison to 20 mm after exiting the die orifice. The post-blow diameter of the molding (molded part) is 140 mm. Compute the corresponding wall thickness of the molding.

Solution

By reference to the solutions of Examples 4.5 and 4.6, we can summarize the data as

$$\text{Swell ratio} = r_s = 2.1$$

The mean diameter of the parison as it exits the die $= D_d = 9.5$ mm.

The pre-blow diameter of the molding $= D_x = 20$ mm.

The wall thickness of the *parison* tube as it exits the extruder die $= t_d = 7.5$ mm.

The wall thickness of the extruded *parison* $= t_x = 15.7$ mm.

The post-blowing diameter of the molding $= D_m = 140$ mm.

The post-blow wall thickness of the molding $= t_m = ?$

By using Equation 4.12,

$$t_m = (r_s^2 t_d D_d)/D_m = [(2.1)^2 \times 7.5 \times 9.5]/140 = 2.24 \text{ mm}$$

Hence, the post-blow wall thickness of the molding $= 2.24$ mm.

EXAMPLE 4.8: COMPUTING THE DIMENSIONS OF MOLD CAVITY IN AN INJECTION MOLDING MACHINE

The nominal length of an injection molded part made of PVC is to be 80 mm. Compute the corresponding dimensions of the mold cavity that will compensate for shrinkage during cooling of the PVC plastic.

Solution

$D_m = 80$ mm; $S = 0.005$ mm/mm (see Table 4.1).

By using Equation 4.17,

$$D_c = D_m + (D_m S) + (D_m S^2) = 80 + (80 \times 0.005) + [(80) \cdot (0.005)^2]$$
$$= 80.402 \text{ mm}$$

The dimensions of the mold cavity for the injection molding process $= 80.402$ mm.

EXAMPLE 4.9: DETERMINATION OF THE COOLING TIME FOR INJECTION MOLDING

The mold wall temperature, in an injection molding machine, is assumed to be 50°C. The temperature of the molten ABS plastic, during the injection molding process, is 230°C; the molded product ejection temperature is 90°C. The thermal diffusivity of ABS plastic at 50°C is 0.0827 mm²/s.

Determine the minimum cooling time for the injection molded ABS plastic product with a wall thickness of 1.5 mm.

Solution

Wall thickness of the molded product $= h = 1.5$ mm.

Thermal diffusivity of the plastic at the mold wall temperature $= \alpha = 0.0827$ mm^2/s.

The temperature of the molten plastic $= T_M = 230°$C.

The mold wall temperature $= T_W = 50°$C. The ejection temperature $= T_E = 90°$C.

By using Equation 4.18,

$$t_c = (h^2/\alpha \cdot \pi^2) \{\ln [(4T_M - 4T_W)/(\pi T_E - \pi T_W)]\}$$

$$t_c = [1.5^2/(0.0827 \cdot 9.87)] \{\ln [(4 \cdot 230 - 4 \cdot 50)/(\pi \cdot 90 - \pi \cdot 50)]\} = 4.8 \text{ s}$$

The minimum cooling time required until the average temperature of the injection molded product reaches the mold wall temperature $= 4.8$ seconds.

EXAMPLE 4.10: CALCULATING THE CLAMPING FORCE FOR INJECTION MOLDING

The pressure inside the cavity of an injection molding machine is 240 kgf/cm^2. The projection area of one cavity is 15.3 cm^2, and the projection area of the runner is 5.5 cm^2. Calculate the required mold clamping force when four molded plastic parts are produced.

Solution

$p = 240$ kgf/cm^2, Total projection area $= A = 15.3 + (4 \times 5.5) = 37.3$ cm^2.

By using Equation 4.19,

$$F = (pA)/1000 = (240 \times 37.3)/1000 = 8.95 \text{ tf}$$

Hence, an injection molding machine having a mold clamping force of 10 tf is required. Giving some margin, it is optimum to select an injection molding machine with a 15 tf rating.

EXAMPLE 4.11: COMPUTING THE EXTRUSION DIE DIAMETER FOR BLOWN FILM EXTRUSION PROCESS

A typical polyethylene shopping bag, manufactured by the blown film extrusion process, has a lateral dimension (width) of 350 mm. Compute the extrusion die diameter.

Solution

By using Equation 4.16,

$$d_d = (2w)/(2.5\pi) = (2 \times 350)/(2.5 \times 3.142) = 89.1 \text{ mm}$$

Hence, the extrusion die diameter (for the blown film extrusion machine) = 89.1 mm.

EXAMPLE 4.12: DETERMINING THE SOLIDS CONVEYING RATE IN FEED ZONE OF AN EXTRUDER

Determine the solids conveying rate in the feed zone of an extruder screw. The geometric design parameters of the feeding zone of the screw are as follows:

Screw diameter = 30 mm, screw pitch = 30 mm, number of flights = 1, flight width = 3 mm, channel width = 28.6 mm, depth of the feed zone = 5 mm, conveying efficiency = 0.436, screw speed = 250 rpm, bulk density of the polymer = 800 kg/m³.

Solution

$D = 30 \text{ mm} = 0.03 \text{ m}$, $p = 30 \text{ mm} = 0.03 \text{ m}$, $w_{FLT} = 3 \text{ mm} = 0.003 \text{ m}$, $W = 28.6 \text{ mm} = 0.0286 \text{ m}$, $H = 5 \text{ mm} = 0.005 \text{ m}$, $\eta_F = 0.436$, $N = 250 \text{ rpm}$, $\rho = 800 \text{ kg/m}^3$.

Since $D = p$, it is a square screw. By reference to the solution of Example 4.1, $\phi = 17.6°$.

By using Equation 4.20,

$$G = 60\rho N \eta_F \pi^2 HD(D - H) \, [W/(W + w_{FLT})](\sin \phi) (\cos \phi)$$

$$G = 60 \times 800 \times 250 \times 0.436 \times 0.005 \times 0.03 \, (0.03 - 0.005)$$
$$[(0.0286)/(0.028 + 0.003)] \, (\sin 17.6°) (\cos 17.6°) \, G = 50 \text{ kg/h}$$

Solids conveying rate = $G = 50$ kg per hour.

EXAMPLE 4.13: DETERMINING THE PISTON SPEED IN INJECTION MOLDING

The oil volume flow rate in an injection molding machine pump is given at 55 gal/min. The piston cross-sectional area is 44.2 in². Calculate the *speed of piston travel*.

Solution

Oil volume flow rate = $Q_o = 55 \text{ gal/min} = 55 \times (1728/7.48)$ in³/min = 12,706 in³/min.

Piston cross-sectional area = $A_p = 44.2$ in².

By using Equation 4.21

$$v_p = Q_o/(60A_p) = (12{,}706)/(60 \times 44.2) = 4.8 \text{ in/s}$$

The *speed of piston travel* $= 4.8$ in/s.

Questions and Problems

4.1 Explain the term *die swell* with the aid of a diagram.

4.2 Differentiate between extrusion molding and injection molding machines with the aid of sketches.

4.3 (a) Sketch the model of drag flow of fluid in the barrel of an extrusion molding machine. (b) Derive an expression for the volume drag flow rate of fluid, Q_d, in terms of screw flight diameter, screw rotational speed, screw helix angle, and the screw channel depth.

4.4 Explain the process used to manufacture thin-walled plastic bottles with the aid of diagrams.

P4.5 The extrusion die, for plastic parison used in blow molding, has a mean diameter of 19 mm. The diameter of the ring opening in the extrusion die is 2 mm. The die swell results in an increase of the mean diameter of the parison to 23 mm after exiting the die orifice. Compute the swell ratio.

P4.6 An extrusion molding machine's barrel has a diameter of 10 cm. The screw rotates at 1.4 rev/s. The channel depth is 7.5 mm, and the screw flight angle is 19°. Calculate the *volume drag flow rate* of the plastic in the barrel of the extrusion machine.

P4.7 An extruder barrel has a diameter of 9 cm. The screw rotates at 1.1 rev/s. The screw flight angle is 20°. Compute the velocity of the screw surface in the stationary barrel.

P4.8 Determine the solids conveying rate in the feed zone of an extruder screw. The geometric design parameters of the feeding zone of the screw are as follows: screw diameter $= 28$ mm, screw pitch $= 28$ mm, number of flights $= 1$, flight width $= 3.2$ mm, channel width $= 29$ mm, depth of the feed zone $= 4.8$ mm, conveying efficiency $= 0.436$, screw speed $= 270$ rpm, and bulk density of the polymer $= 800$ kg/m^3.

P4.9 An extruder barrel has a diameter of 8.8 cm. The screw rotates at 1.3 rev/s. The screw flight angle is 17°. Calculate the flight width of the extruder screw.

P4.10 The extrusion die, for plastic parison used in blow molding, has a mean diameter of 17.5 mm. The diameter of the ring opening in the die is 2.0 mm. The die swell results in an increase of the mean diameter of the parison to 21 mm after exiting the die orifice. The post-blow diameter of the molding (molded part) is 145 mm. Compute the corresponding wall thickness of the molding.

P4.11   The pressure inside the mold cavity of an injection molding machine is $270 \, kgf/cm^2$. The projection area of one cavity is $16 \, cm^2$, and the projection area of the runner is $5.8 \, cm^2$. Calculate the required mold clamping force when two molded plastic parts are produced.

References

Isayev, A.I. (Ed.). 2016. *Advances in Polymer Technology*, 35(4). New York: Wiley.

Kalpakjian, S., Schmid, S. 2013. *Manufacturing Engineering and Technology*, 7th Edition. New York: Pearson.

Kazmer, D. 2007. *Injection Mold Design Engineering*. Munich: Hanser.

Kimerling, T. 2002. Injection Molding Cooling Time Reduction and Thermal Stress Analysis. Internet Source: http://www-unix.ecs.umass.edu/mie/labs/mda/fea/fealib/Tom%20Kimerling/TKimerling_injection_molding.pdf.

Koopmans, R.J. 1999. Die swell or extrudate swell. In *Polypropylene: An A-Z Reference* (Ed: Karger-Kocsis), pp. 158–162. Dordrecht: Kluwer Publishers.

Mitani, K. 1997. *Molds for Injection Molding (in Japanese)*. Japan: Sigma.

Rao, N.S., Schott, N.R. 2012. *Understanding Plastics Engineering Calculations: Hands-on Examples and Case Studies*. Munich: Hanser.

Rauwendaal, S. 2014. *Polymer Extrusion*, 5th Edition. Munich: Hanser.

5

Metal Forming I—Deformation and Annealing

5.1 Bulk Deformation and Sheet Metal Forming

All metal forming processes may be classified into two groups: (1) bulk deformation and (2) sheet metal forming. Bulk deformation processes include rolling, forging, extruding, bar and wire drawing, and the like. Sheet metal forming processes include bending, shearing (and blanking and piercing), deep drawing/cup drawing, and the like.

Bulk deformation involves deformation with a significant change in the thickness or cross-sectional area of the workpiece. *Sheet metal forming* processes involve deformation with no or little change in the cross-sectional area of the workpiece. Bulk deformation and sheet metal forming processes are discussed in Chapters 6 and 7, respectively.

5.2 Deformation of Single Crystal and Its Mathematical Modeling

Deformation in ductile materials occurs by shear along a *slip system*. The differences in the deformation behavior of metals having different crystal structures can be explained by examining the force required to initiate the slip process. Consider a unidirectional tensile force, F, applied to a cylinder of metal that is a single crystal (see Figure 5.1).

In order to deform the metal (i.e., to move the dislocation in its slip system), a shear force acting in the slip direction must be produced by the applied force. This resolved shear force, F_r, can be given by (see Figure 5.1)

$$F_r = F \cos \lambda \tag{5.1}$$

where λ is the angle between the applied force, F, and the slip direction.

By dividing both sides of Equation 5.1 by A, the area of slip plane, we obtain

$$(F_r/A) = (F/A) \cos \lambda \tag{5.2}$$

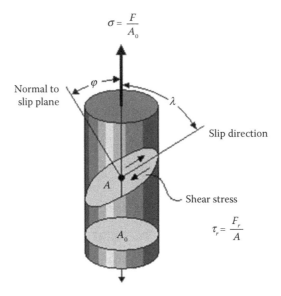

FIGURE 5.1
A resolved shear stress, τ_r, that is produced on a slip system.

By reference to Figure 5.2, $F_r/A = \tau_r$, $A = A_0/\cos\varphi$, and $\sigma = F/A_0$; these relationships lead us to rewrite Equation 5.2 as follows:

$$\tau_r = \sigma(\cos\varphi)(\cos\lambda) \tag{5.3}$$

where τ_r is the resolved shear stress; σ is the tensile stress; φ is the angle between the applied force (F) and normal to the slip plane; and λ is the angle between F and the slip direction. The mathematical model, presented in Equation 5.3, is called *Schimd's law*, and it enables us to compute the resolved shear stress when the other variables are known in a single crystal.

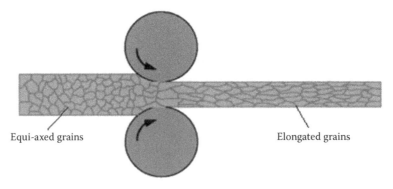

FIGURE 5.2
Elongation of grains due to CW of metal.

5.3 Cold Working and Hot Working

Cold forming or *cold working* (CW) involves metal forming at a temperature below its recrystallization temperature. *Hot forming* or *hot working* (HW) involves metal forming at a temperature above the recrystallization temperature of the metal. An example of CW is the forming or working of aluminum at room temperature. Steels are usually hot worked (HW) at a temperature in the range of 1200°C–1280°C.

5.4 Cold Work and Strain Hardening

When a metal is CW, dislocations moving through the material entangle with one another, thereby causing dislocation motion to become more difficult in the CW material (Guo, 2005). This metallurgical phenomenon is known as *strain hardening*. Strain hardening or *work hardening* means material becomes harder and stronger but less ductile due to CW.

Strain hardening in CW metals can be mathematically represented by

$$\sigma_T = K(\varepsilon_T)^n \tag{5.4}$$

where σ_T is the true stress (yield stress or flow stress), K is the strength coefficient, ε_T is the true strain, and n is the work-hardening exponent. A value of $n = 0$ means that a material is a perfectly plastic solid, while a value of $n = 1$ represents a 100% elastic behavior of the solid. Hence, in order to strengthen a metal by strain hardening, an n value in the range of 0.10–0.50 is usually recommended. Table 5.1 lists the n and K values of commonly used metallic materials.

The degree of CW is usually expressed as %CW, which is given by

$$\%CW = [(A_0 - A_d)/A_0] \times 100 \tag{5.5}$$

TABLE 5.1

Strength Coefficient (K) and Work-Hardening Exponents (n) Data

Materials	Copper (annealed)	70–30 Brass (annealed)	0.05%C Steel (annealed)	0.6%C Steel (Q&T at 700°C)	SAE 4340 Steel	2024 Al Alloy (H/T)
n	0.54	0.49	0.22	0.19	0.15	0.17
K, MPa	320	900	600	1230	641	600

Note:
Q&T = quenched and tempered.
H/T = heat treated.

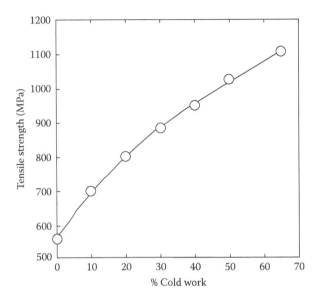

FIGURE 5.3
Dependence of tensile strength on the %CW in austenitic stainless steel.

where A_0 is the original cross-sectional area, and A_d is the deformed cross-sectional area.

When a metal is CW by cold rolling, the grains are elongated in the rolling direction as shown in Figure 5.2.

The tensile strength of a CW metal strongly depends on the degree of cold work (%CW). Figure 5.3 illustrates the variation of the tensile strength with %CW for austenitic stainless steel; this material behavior is, obviously, due to strain hardening.

5.5 Annealing: Recovery, Recrystallization, and Grain Growth

It is learned in the preceding section that CW results in an increase in hardness and reduction in ductility. In order to restore ductility, the CW metal is usually annealed (heated); this additional heat treatment is called *annealing*. There are three stages of annealing: (1) recovery, (2) recrystallization, and (3) grain growth (Huda, 2017). *Recovery* involves the relief of internal stresses due to CW. In order to design and use forming processes, one must deeply study the recrystallization and grain growth behaviors of the metal. The grain size and the mechanical properties of an annealed metal strongly depend on the annealing temperature; this dependence is illustrated in Figure 5.4.

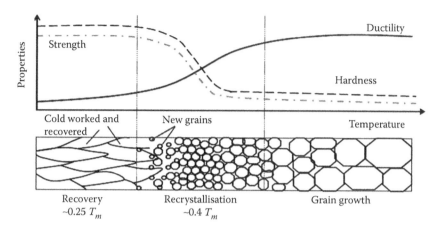

FIGURE 5.4
The effects of annealing temperature on microstructure and mechanical properties of metal.

5.5.1 Recrystallization Annealing

It is the recrystallization stage of annealing that mainly restores the ductility of CW metal. Recrystallization refers to the rearrangement and reorientation of the elongated CW grains to the equi-axed grains when a metal is heated at its recrystallization temperature, T_r. Recrystallization temperature, T_r, is the minimum temperature at which complete recrystallization occurs in an annealed CW metal within a specified time.

The recrystallization temperature, T_r, is related to melting temperature, T_m of a metal by

$$T_r \sim 0.4\ T_m \tag{5.6}$$

where the temperatures are measured in Kelvin (K).

Recrystallization is a time-dependent process; the time dependence of the fraction of statically recrystallized material is commonly described through the Avrami general equation (Callister, 2007; Tajjaly and Huda, 2011):

$$X = 1 - \exp(-kt^n) \tag{5.7}$$

where X is the fraction of recrystallization; t is the annealing time; and k and n are time-independent constants for the particular process. The kinetics of static recrystallization are often characterized in terms of the time required for recrystallization of 50% of the material (i.e., $t_{0.5}$).

The rate of recrystallization can be computed as the reciprocal of the time required for recrystallization of 50% of the material (i.e., $t_{0.5}$) as follows:

$$\text{Rate} = (1/t_{0.5}) \tag{5.8}$$

The determination of fraction recrystallized usually involves a metallographic technique, which consists of quenching the material and mapping the locations of grains, their sizes, and their shapes in an optical microscope to find the ratio of recrystallized grains. This is a complicated technique. Kliber and co-researchers (2012) reported a convenient stress-relaxation testing method to compute the recrystallization fraction. This testing method involves heating to a required temperature, cooling to deformation temperature, deformation at a prescribed rate, and then holding the strain constant while recording the force (or stress) and time (from the start of the hold) (Kliber et al., 2012). By using the stress-relaxation test, it is possible to calculate the recrystallization fraction, X, by the following formula:

$$X = \{[\sigma_{01} - (\alpha_1 \log t)] - \sigma\}/\{(\sigma_{01} - \sigma_{03}) - (\alpha_1 - \alpha_3) \log t\} \qquad (5.9)$$

where α_1 is the slope of the stress-relaxation curve (recovery stage); α_3 is the slope of the curve (final stage); σ_{01} and σ_{03} are the stress values during the recovery and final stage, respectively; σ is the stress value in the second part of the curve, MPa; and t is the relaxation time, s (see Example 5.13).

5.5.2 Grain Growth

Grain growth is the third stage of annealing; it refers to the increase in the average grain size of the metal when the metal is heated beyond the completion of recrystallization (see Figure 5.4). The control of grain growth is of great technological importance because many material properties strongly depend on the grain size and its distribution (Huda and Ralph, 1990; Huda et al., 2014).

Many polycrystalline materials follow the following behavior during grain growth (Huda, 2004):

$$D^m - D_0^m = K t \qquad (5.10)$$

where D is the average grain diameter at the end of isothermal annealing at a specified temperature for a time duration of t; K is a temperature-dependent constant; D_0 is the pregrowth grain diameter; and m is the reciprocal of the grain-growth exponent, n. For most metals, the value of $n = 0.5$ (i.e., $m = 2$) (Huda et al., 2014); thus Equation 5.10 may be rewritten as

$$D^2 - D_0^2 = K t \qquad (5.11)$$

For a polycrystalline material with extremely fine grain size, the original grain diameter may be taken as zero (i.e., $D_0 = 0$). Thus, Equation 5.10 takes the following form:

$$D^m = K t \qquad (5.12)$$

Since m is the reciprocal of the grain-growth exponent, n (i.e., $m = 1/n$), Equation 5.12 becomes

$$D = k\, t^n \tag{5.13}$$

where k is a new constant. This equation (Equation 5.13) is valid only for fine-grained materials.

5.6 Hot Working of Metals

It is learned in Section 5.3 that *HW* occurs at a temperature above the recrystallization of the metal. Metal forming processes are often performed at high speeds that do not allow sufficient time for complete recrystallization of the grain structure during the deformation cycle itself; however, recrystallization may occur immediately following the forming process or later as the metal cools. In general, recrystallization during *HW* not only results in rearrangement of crystals into equi-axed grains but also in *grain refinement*.

A distinct advantage of *HW* over *CW* is that the former process does not need an annealing treatment because recrystallization occurs during the *HW* process.

The degree of hot work (%*HW*) can be expressed by the relationship

$$\%HW = [(A_0 - A_d)/A_0] \times 100 \tag{5.14}$$

where A_0 is the original cross-sectional area, and A_d is deformed cross-sectional area.

5.7 Problems and Solutions—Examples in Deformation and Annealing

EXAMPLE 5.1: COMPUTING THE RESOLVED SHEAR STRESS FOR SINGLE-CRYSTAL METALLIC MATERIAL

A metal single-crystal specimen, having a diameter of 6 mm, starts to deform plastically at a tensile load of 9000 N. Compute the critical resolved shear stress, given that the angle between the slip direction and tensile axis is 55° and the normal to slip plane is at 45° to the tensile axis.

Solution

$\lambda = 55°$, $\varphi = 45°$, Tensile force $= F = 9000\,\text{N}$, $\tau_r = ?$

$$\text{Area} = A_0 = (\pi/4)\, d^2 = (\pi/4)\, 6^2 = 28.27\,\text{mm}^2$$

The tensile stress, σ, can be calculated as follows:

$$\sigma = F/A_0 = 9000/28.27 = 318.3 \text{ MPa}$$

Now, by applying Schimd's law (Equation 5.3),

$$\tau_r = \sigma \, (\cos \varphi) \, (\cos \lambda) = 318.3 \, (\cos 45°) \, (\cos 55°) = 129 \text{ MPa}$$

Hence, the critical resolved shear stress is 129 MPa.

EXAMPLE 5.2: CALCULATING THE DIAMETER OF STOCK MATERIALS WHEN %CW IS GIVEN

It is required to obtain a diameter of 5.1 mm in a cold-formed (cold drawn) metal bar. The strength requirement demands 21.5%CW during the bar drawing. What should be the diameter of the stock material (bar)?

Solution

First, we can find the deformed area of cross section as follows:

$$A_d = (\pi/4) \, d^2 = (\pi/4) \, (5.1)^2 = 20.4 \text{ mm}^2$$

By using Equation 5.5,

$$\%CW = [(A_0 - A_d)/A_0] \times 100$$

$$21.5 = [(A_0 - 20.4)/A_0] \times 100$$

$$A_0 = 26 \text{ mm}^2$$

Now, we can find the original diameter, d_0, as follows:

$$A_0 = (\pi/4) \, d_0^2$$

$$26 = (\pi/4) \, d_0^2; \quad d_0 = 5.75 \text{ mm}$$

Hence, the diameter of the stock material (bar) should be 5.8 mm.

EXAMPLE 5.3: ESTIMATING TENSILE STRENGTH WHEN DIMENSIONS BEFORE AND AFTER CW ARE KNOWN

A sample of austenitic stainless steel having cross-sectional area of 30 mm² was CW to deformed cross-sectional area of 18 mm². What is the expected tensile strength of the deformed steel?

Solution

$A_0 = 30 \text{ mm}^2$; $A_d = 18 \text{ mm}^2$.

By using Equation 5.5,

$$\%CW = [(A_0 - A_d)/A_0] \times 100 = [(30{-}18)/30] \times 100 = 40$$

$$\%CW = 40$$

By reference to Figure 5.3, $\%CW = 40$ corresponds to a tensile strength of about 930 MPa.

EXAMPLE 5.4: COMPUTING THE WORK-HARDENING EXPONENT

Annealed copper was cold formed by deforming by a true strain of 0.8. What true stress was applied to allow flow of the material for CW?

Solution

$\varepsilon_T = 0.8$, $\sigma_T = $ flow stress $= ?$

By reference to the data for annealed copper in Table 5.1, we see

$n = 0.54$, and $K = 320$ MPa

By using Equation 5.4,

$$\sigma_T = K\,(\varepsilon_T)^n$$
$$\sigma_T = 320\,(0.8)^{0.54} = 320 \times 0.88 = 284 \text{ MPa}$$

Hence, the flow stress $= 284$ MPa.

EXAMPLE 5.5: COMPUTING $\%CW$ WHEN THE WORK-HARDENING EXPONENT IS GIVEN

The stress-strain behavior of annealed 304 stainless steel follows the following relationship:

$$\sigma = 1275\,(\varepsilon_{CW} + 0.002)^{0.45}$$

Calculate the degree of CW that has to be imposed for yield strength of 780 MPa in the metal.

Solution

$\sigma_T = \sigma_{ys} = 780$ MPa

The stress, σ, in the given relationship refers to σ_T in Equation 5.4; thus, we may write

$$780 = 1275\,(\varepsilon_{CW} + 0.002)^{0.45}$$

$$\text{or} \quad 0.611 = (\varepsilon_{CW} + 0.002)^{0.45}$$

$$\text{or} \quad (\varepsilon_{CW} + 0.002) = (0.611)^{1/0.45} = (0.611)^{2.22} = 0.335$$

$$\varepsilon_{CW} = 0.333$$

$$\text{But} \quad \varepsilon_{CW} = \text{true strain} = \ln (A_0/A_i)$$

$$\text{or} \quad A_0/A_{CW} = e^{0.333} = 1.39$$

$$A_{CW}/A_0 = 0.716$$

By using Equation 5.5,

$$\%CW = [(A_0 - A_d)/A_0] \times 100 = [(A_0 - A_{CW})/A_0] \times 100$$
$$= [1 - (A_{CW}/A_0)] \times 100$$

$$\%CW = (1 - 0.716) \times 100 = 28.3$$

Hence, the degree of CW required to be imposed in the steel $= 28.3\%$.

EXAMPLE 5.6: COMPUTING ORIGINAL DIAMETER WHEN THE POST-CW TENSILE STRENGTH IS GIVEN

A cylindrical specimen of CW austenitic stainless steel has a tensile strength of 1000 MPa. If its CW diameter is 8 mm, what was its diameter before deformation?

Solution

Deformed cross-sectional area $= A_d = (\pi/4) D_d^2 = (\pi/4) (8)^2 = 50.26 \text{ mm}^2$.

By reference to Figure 5.3, the tensile strength of 1000 MPa corresponds to $\%CW = 50$.

By using Equation 5.5,

$$\%CW = [(A_0 - A_d)/A_0] \times 100 = [(1 - (A_d/A_0)] \times 100$$
$$50 = [(1 - (50.26/A_0)] \times 100$$

$$A_0 = 100.5 \text{ mm}^2$$

Now, we can find the original diameter, D_0, as follows:

$$A_0 = (\pi/4) D_0^2$$

$$100.5 = (\pi/4) D_0^2$$

$$D_0 = 11.3 \text{ mm}$$

Hence, the diameter of the specimen before deformation should be 11.3 mm.

EXAMPLE 5.7: COMPUTING THE RECRYSTALLIZATION TEMPERATURE

The melting temperature of aluminum is 660°C and that of lead is 327°C. Compute the recrystallization temperatures of the two metals.

Solution

Melting temperature of aluminum $= T_m = 660°C = 660 + 273 = 933\ K$.

By using Equation 5.6,

$$T_r = 0.4\ T_m = 0.4 \times 933 = 373.2\ K = 373.2 - 273 = 100.2°C$$

Recrystallization temperature of aluminum $= 101°C$.

Melting temperature of lead $= T_m = 327°C = 327 + 273 = 600\ K$.

Recrystallization temperature of lead $= T_r = 0.4\ T_m = 0.4 \times 600 = 240\ K = 240 - 273 = -33°C$.

EXAMPLE 5.8: DECIDING WHETHER OR NOT A METAL NEEDS A RECRYSTALLIZATION ANNEALING TREATMENT

Two different samples of aluminum and lead metals were formed at room temperature (25°C). Which of the two metals does not require a furnace for recrystallization annealing? Explain!

Solution

We refer to the data obtained in the solution of Example 5.7. For aluminum, the working temperature (25°C) is below the recrystallization temperature of *Al* (101°C). It means aluminum was *CW* at 25°C; hence, *Al* needs to be recrystallized in a furnace.

For lead, the working temperature (25°C) is above the T_r of lead (−33°C). It means that lead was *HW* at 25°C; hence, lead does *not* need to be recrystallized in a furnace.

EXAMPLE 5.9: APPLICATION OF THE AVRAMI EQUATION TO DETERMINE THE CONSTANT K

The kinetics of recrystallization for some alloy follow the Avrami relationship (Equation 5.7) and that the value of n in the exponential is 2.3. If, after 120 s, 50% recrystallization is complete, compute the constant k.

Solution

For the Avrami relationship, $n = 2.3$, $t = 120\ s$, $X = 50\% = 0.50$, $k = ?$

By using Equation 5.7,

$$X = 1 - \exp(-kt^n)$$

$$\text{or} \quad \exp(-kt^n) = 1 - X$$

$$\text{or} \quad -kt^n = \ln (1 - X)$$

$$\text{or} \quad k = -\{[\ln (1 - X)]/t^n\}$$

$$k = -\{[\ln (1 - 0.5)]/120^{2.3}\} = -\{[\ln (0.5)]/60550.3\} = 1.14 \times 10^{-5}$$

$$k = 1.14 \times 10^{-5}$$

EXAMPLE 5.10: APPLICATION OF THE AVRAMI EQUATION TO DETERMINE THE RECRYSTALLIZATION TIME

Using the data in Example 5.9, determine the total time to complete 95% recrystallization.

Solution

For the Avrami relationship, $n = 2.3$, $X = 95\% = 0.95$, $k = 1.14 \times 10^{-5}$, $t = ?$

By using Equation 5.7,

$$X = 1 - \exp(-kt^n)$$

$$\text{or} \quad t^n = -\{[\ln (1 - X)]/k\}$$

$$t = \{-[\ln (1 - X)]/k\}^{1/n}$$

$$t = \{-[\ln (1 - 0.95)]/(1.14 \times 10^{-5})\}^{1/2.3} = \{-[\ln (0.05)]/(1.14 \times 10^{-5})\}^{0.434}$$

$$t = \{-[(-2.62 \times 10^5)]\}^{0.434} = (2.62)^{0.434} \times 10^{2.17} = 1.52 \times 148 = 225 \text{ s}$$

The total time to complete 95% recrystallization $= 225 \text{ s} = 3.74$ min.

EXAMPLE 5.11: CALCULATING THE RATE OF RECRYSTALLIZATION

Determine the rate of recrystallization of the alloy with the parameters stated in Example 5.9.

Solution

In Example 5.9, time to complete 50% recrystallization $= t_{0.5} = 120$ s.

By using Equation 5.8,

$$\text{Rate} = (1/t_{0.5}) = 1/120 = 0.008/\text{s}$$

The rate of recrystallization of the alloy $= 0.008 \text{ s}^{-1}$.

EXAMPLE 5.12: COMPUTING THE FRACTION OF RECRYSTALLIZATION WHEN N AND K ARE UNKNOWN

A CW aluminum sample was recrystallized at $350°C$ for 95 min; this annealing treatment resulted in 30% recrystallized microstructure. Another sample of the same CW Al metal was recrystallized for 127 min at the same temperature, which resulted in 80% recrystallized structure. Determine the fraction recrystallized after an annealing time of 110 min, assuming that the kinetics of this recrystallization process obey the Avrami equation.

Solution

By using Equation 5.7,

$$X = 1 - \exp(-kt^n)$$

$$\text{or} \quad k = -\{[\ln (1 - X)]/t^n\}$$

$X = 0.3$ when $t = 95$ min

$$k = -\{[\ln (1 - 0.3)]/95^n\}$$

$$k = 0.35/95^n \tag{E5.12.1}$$

$X = 0.8$ when $t = 110$ min.

$$k = -\{[\ln (1 - 0.8)]/110^n\}$$

$$k = 1.6/127^n \tag{E5.12.2}$$

By solving Equations (E5.12.1) and (E5.12.2) simultaneously, we obtain $n = 5.2, k = 1.8 \times 10^{-11}$

In order to determine the fraction recrystallized after an annealing time, $t = 110$ min,

$$X = 1 - \exp(-kt^n) = 1 - \{\exp[-(1.8 \times 10^{-11}) \, 110^{5.2})]\}$$
$$= 1 - [\exp(-0.74)] = 1 - 0.47 = 0.52$$

$$X = 0.52$$

The fraction recrystallized after an annealing time of 110 min $= 0.52$.

EXAMPLE 5.13: COMPUTING STRESS FOR RECRYSTALLIZED FRACTION BASED ON STRESS-RELAXATION TEST

A metal sample was recrystallized by using a stress-relaxation testing method. Calculate the stress value in the second part of the stress-relaxation curve based on the following data:

The slope of the stress-relaxation curve (recovery stage) = 25 MPa.

The slope of the stress-relaxation curve (final stage) = 8 MPa.

Stress during the recovery stage = 65 MPa.

Stress during the final stage = 20 MPa.

Time to complete 50% recrystallization = 2.3 s; $n = 1.3$.

Solution

$\alpha_1 = 25$ MPa, $\alpha_3 = 8$ MPa, $\sigma_{01} = 65$ MPa, $\sigma_{03} = 20$ MPa, $n = 1.3$, $X = 0.5$, $t_{0.5} = 2.3$ s, $\sigma = ?$

By using Equation 5.9,

$$X = \{[\sigma_{01} - (\alpha_1 \log t)] - \sigma\}/\{(\sigma_{01} - \sigma_{03}) - (\alpha_1 - \alpha_3) \log t\}$$

$$0.5 = \{[65 - (25 \log (2.3))] - \alpha\}/\{(65 - 20) - (25 - 8) \log (2.3)\}$$

$$0.5 = \{[65 - (25 \times 0.36)] - \alpha\}/\{(45) - (17 \times 0.36)\}$$

$$\sigma = 36.5 \text{ MPa}$$

The stress value in the second part of the stress-relaxation curve = 36.5 MPa.

EXAMPLE 5.14: COMPUTING ANNEALING TIME FOR GRAIN GROWTH PROCESS FOR A METAL

It takes 3000 minutes to increase the average grain diameter of brass from 0.03 to 0.3 mm at 600°C. Assuming that the grain-growth exponent of brass is equal to 0.5, calculate the time required to increase the average grain diameter of brass from 0.3 to 0.6 mm at 600°C.

Solution

For $D_0 = 0.03$ mm and $D = 0.3$ mm, $t = 3000$ min.

By using Equation 5.11,

$$D^2 - D_0^2 = K t$$

$$(0.3)^2 - (0.03)^2 = K(3000)$$

$$K = 2.97 \times 10^{-5}$$

Now, we can compute t for $D_0 = 0.3$ mm, and $D = 0.6$ mm; $K = 2.9 \times 10^{-5}$ (since the temperature is the same) by using Equation 5.11 as follows:

$$(0.6)^2 - (0.3)^2 = (2.9 \times 10^{-5}) t$$

$$t = 9000 \text{ min}$$

Hence, the time required to increase the average grain diameter of brass from 0.3 to 0.6 mm at $600°C = 9000$ minutes.

EXAMPLE 5.15: COMPUTING ANNEALING TIME WHEN THE GRAIN-GROWTH EXPONENT IS UNKNOWN

An extremely fine-grained material was annealed for 2000 minutes to obtain an average grain diameter of 0.02 mm at temperature T. In another experiment, the same fine-grained material was annealed at T for 6000 minutes and the resulting average grain diameter was 0.03 mm.

Compute the following: (a) the grain-growth exponent for the material and (b) the annealing time to obtain an average grain diameter of 0.04 mm at the same temperature T.

Solution

Since this problem deals with a fine-grained material ($D_0 = 0$), Equation 5.13 is applicable.

(a) By using Equation 5.13 for the first experiment ($D = 0.02$ mm, $t = 2000$ min),

$$D = k\, t^n,$$

$$0.02 = k\, (2000)^n \qquad\qquad (E5.15.1)$$

By using Equation 5.13 for the second experiment ($D = 0.04$ mm, $t = 6000$ min),

$$0.03 = k\, (6000)^n \qquad\qquad (E5.15.2)$$

By solving the two equations simultaneously, we get

$$n = 0.37, \text{ and } k = 0.0012$$

Hence, the grain-growth exponent of the material $= 0.37$.

(b) In order to compute the annealing time for $D = 0.04$ mm, we again use Equation 5.13:

$$D = k\, t^n$$

$$0.04 = 0.0012\, t^{0.37}$$

$$\log 0.04 = \log (0.0012) + (0.37) \log t$$

$$\log t = 4.13, \quad t = 10^{4.13} = 13{,}490$$

The required annealing time $= t = 13{,}490$ min.

EXAMPLE 5.16: CALCULATING PERCENT *HW*

A metallic strip having cross-sectional area of 45 mm^2 was *HW* to a deformed cross-sectional area of 30 mm^2. Calculate the *percent hot work* (%*HW*).

Solution

$A_0 = 45 \text{ mm}^2$, $A_d = 30 \text{ mm}^2$.

By using Equation 5.14,

$$\%HW = [(A_0 - A_d)/A_0] \times 100 = [(45-30)/45] \times 100 = 33.3$$

Hot Work $= 33.3\%$.

Questions and Problems

5.1 (a) Differentiate between *CW* and *HW*. (b) Draw a classification chart showing the various types of metal forming processes.

5.2 (a) What are the meanings of the work-hardening exponent values: (i) $n = 0$, (ii) $n = 1$. (b) Why do *CW* metals generally require recrystallization annealing?

5.3 Explain the effects of annealing temperature on the microstructure and mechanical properties of a metal with the aid of a diagram.

5.4 Compare the metallographic technique of determining the fraction recrystallized with the stress-relation technique.

5.5 Show that grain growth behavior in a fine-grained material may be expressed as $D = kt^n$.

5.6 Why do *HW* metals generally not require recrystallization annealing treatment?

P5.7 A metal single-crystal specimen starts to deform plastically at a tensile stress of 300 MPa. Compute the critical resolved shear stress, given that the angle between the slip direction and tensile axis is 53° and the normal to slip plane is at 45° to the tensile axis.

P5.8 A sample of austenitic stainless steel having cross-sectional area 30 mm² was *CW* to deformed cross-sectional area of 18 mm². What is the expected tensile strength of the deformed steel?

P5.9 The melting temperature of tin is 232°C and that of copper is 1080°C. Compute the recrystallization temperatures of the two metals.

P5.10 A *CW* aluminum alloy was recrystallized at 370°C for 100 min; this annealing treatment resulted in 35% recrystallized microstructure. Another sample of the same *CW* Al metal was recrystallized for 120 min at the same temperature, which resulted in 75% recrystallized structure. Determine the fraction recrystallized after an annealing time (total time) of 140 min. (Assume that the kinetics of this recrystallization process obey the Avrami equation.)

P5.11 A fine-grained material was annealed for 1500 minutes to obtain an average grain diameter of 0.010 mm at temperature *T*. Another sample of the same material was annealed at *T* for 5000 minutes, and the resulting average grain diameter was 0.017 mm. Compute the grain-growth exponent.

References

Callister, W.D. 2007. *Materials Science and Engineering: An Introduction*. New York: John Wiley and Sons.

Guo, Z.X. (Ed). 2005. *The Deformation and Processing of Structural Materials*. Boca Raton, FL: CRC Press/Taylor and Francis.

Huda, Z. 2004. Influence of particle mechanisms on the kinetics of grain growth in a P/M superalloy. *Materials Science Forum*, 467–470, 985–990. http://www.scientific.net/MSF.467-470.985.

Huda, Z. 2017. *Materials Processing for Engineering Manufacture*. Switzerland: Trans Tech.

Huda, Z., Ralph, B. 1990. Kinetics of grain growth in powder formed IN-792 superalloy. *Materials Characterization*, 25, 211–220.

Huda, Z., Zaharinie, T., Metselaar, I.H.S.C., Ibrahim, S., Min, G.J. 2014. Kinetics of grain growth in 718 Ni-base Superalloy. *Archives of Metallurgy and Materials*, 59(3), 847–852.

Kliber, J., Horsinka, J., Knapiński, M. 2014. The static recrystallization kinetics acquired from stress relaxation method. *Hutnik, Wiadomości Hutnicze*, 81(1), 35–38.

Tajjaly, M., Huda, Z. 2011. Recrystallization kinetics for 7075 aluminum alloy. *Metal Science and Heat Treatment*, 53(5–6), 213–217.

6

Metal Forming II—Bulk Deformation

6.1 Metal Rolling

6.1.1 Rolling and Its Types

Rolling is the most economical and the most commonly used metal forming process. *Rolling* involves squeezing a metallic workpiece between two rolls, which exert compressive stresses, thereby reducing the thickness or cross-sectional area of the workpiece (see Figure 1.2).

Rolled steel products include plates, sheets, bars, structural shapes (I-beams, channels, etc.), and rails as well as intermediate shapes for wire drawing or forging (Ginzburg, 1989). Flat rolling is used to reduce the thickness of a rectangular cross section (such as a slab). In "shape rolling," a square cross section (bloom or billet) is formed into a shape such as an I-beam by using grooved rolls. The equipment used in rolling is called a *rolling mill*. Depending on the number of rolls used, rolling mills may be divided into the following types: (1) two-high rolling mill, (2) three-high mill, (3) four-high mill, (4) cluster mill, and (5) tandem rolling mill. Four-high mills use smaller-diameter rolls as compared to three-high mills, and the latter use smaller-diameter rolls as compared to two-high mills (Ginzburg and Ballas, 2000; Huda, 2017).

6.1.2 Mathematical Modeling of Rolling

Figure 6.1 illustrates an analysis of *flat rolling* (i.e., rolling of a metal strip with rectangular cross sections in which the width is greater than the thickness). The strip is reduced in thickness from t_0 to t_f with width of the strip assumed to remain constant during rolling.

Since volume remains constant, the following relationship holds good:

$$t_0 w_0 L_0 = t_f w_f L_f \qquad (6.1)$$

where t_0 is the starting thickness, w_0 is the starting width, L_0 is the strip length before rolling, t_f is the final thickness, w_f is the final width, and L_f is the strip length after rolling.

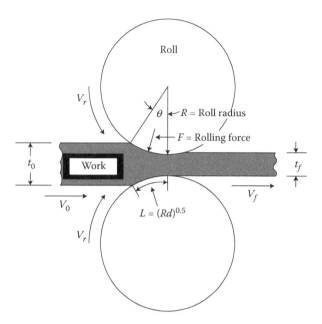

FIGURE 6.1
Analysis of flat rolling of metallic work.

The amount of thickness reduction is expressed as draft, d, as follows:

$$d = t_0 - t_f \tag{6.2}$$

The amount of deformation in rolling is measured by percent reduction, as follows:

$$\% \text{ reduction} = (d/t_0) \times 100 \tag{6.3}$$

where d is draft, mm; and t_0 is the original thickness of the strip, mm.

The forward slip, S, is defined as the difference between the velocity of strip at exit and the roll velocity divided by roll velocity, which is given by (Lu, 2007)

$$S = (v_f - v_r)/v_r \tag{6.4}$$

where v_f is the strip velocity at exit, m/s; and v_r is the roll velocity, m/s. At roll exit, the forward slip is positive, which means that the workpiece moves faster than roll here.

True strain, ε, in the work (strip) is given by

$$\varepsilon = \ln{(t_0/t_f)} \tag{6.5}$$

The contact length, L, is related to the roll diameter, R, by the following relationship (see Figure 6.3):

$$L = (Rd)^{1/2} \tag{6.6}$$

Rolling Force: *Rolling force* is the load with which the rolls press against the work (see Figure 6.1) and is given by

$$F_R = \overline{Y}_f w L \tag{6.7}$$

where F_R is the rolling force, N; $\overline{Y}_f$ is the average flow stress, MPa; and w is work width, mm.

In case of cold working (cold rolling), the average flow stress can be determined by the formula

$$\overline{Y}_f = (K\varepsilon^n)/(1 + n) \tag{6.8}$$

where K is the strength coefficient, MPa, and n is the strain-hardening exponent.

By substituting the value of L from Equation 6.6 into Equation 6.7, we obtain

$$F_R = \overline{Y}_f w (Rd)^{1/2} \tag{6.9}$$

It is evident from Equation 6.9 that the rolling force, F_R, varies directly as the square root of roll radius. It means that the smaller the roll radius, the lesser will be the force required for rolling. This manufacturing principle is applied to the developments of three-high and four-high rolling mills with smaller roll radii (see Section 6.1.1).

Rolling Feasibility: The success or feasibility of rolling strongly depends on the friction between the surface of the roll and the work in contact with the roll. The maximum draft is related to the coefficient of friction by the relationship

$$d_{max} = \mu^2 R \tag{6.10}$$

where d_{max} is the maximum draft, mm; μ is the coefficient of friction between the surfaces along the contact length (see Figure 6.1); and R is the roll radius, mm.

In order for the rolling to be feasible (successful), the following condition must be satisfied:

$$d_{max} > d \tag{6.11}$$

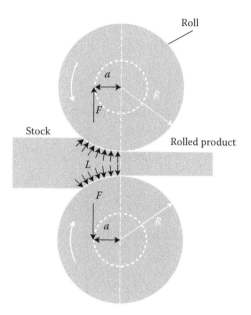

FIGURE 6.2
Basic mechanics applied to flat rolling (F = Force; a = moment arm).

It is evident from Equations 6.10 and 6.11 that larger roll radius (R) would result in a higher value of d_{max}, thereby ensuring success (feasibility) of rolling.

Torque and Power in Rolling: Figure 6.2 shows the basic mechanics applied to flat rolling.

Torque is defined as the vector product of force and moment arm; thus, the magnitude of torque for one roll, T, can be written as

$$T = Fa \approx F(0.5\,L) = 0.5\,FL \qquad (6.12)$$

where T is the rolling torque, N-m; F is rolling force, N; and L is contact length, m.

Since two rolls are necessary for rolling, rolling torque will be

$$\text{Rolling torque} = 2T = FL \qquad (6.13)$$

Power required for rolling can be determined as follows:

$$P = T\omega = (FL)(2\pi N) = 2\pi NFL \qquad (6.14)$$

where P is power, W; N is the roll rotational speed, rev/s; F is rolling force, N; and L is the contact length, m. Equation 6.14 clearly indicates that *power* varies directly as *rolling force*. Since four-high mills and three-high mills require smaller rolling force as compared to two-high mills, the former types of rolling are more energy efficient as compared to the latter type.

6.2 Metal Forging

6.2.1 Forging and Its Types

Metal forging involves plastic deformation of a metal through local compressive forces applied through dies. The forging operation refines grains and results in a fibrous grained structure, which enhances strength. Many high-strength automotive and aerospace components are manufactured by forging. Other examples of forged components include spanner, crane hooks, and the like.

A variety of forging operations are practiced in industry (Ares, 2006). Based on working temperature, forging may be done either by hot working (HW) or by cold working (CW). Based on equipment used, there are two types of forging: drop-hammer forging and press forging. Based on forging operation, there are three types of forging operations: (1) open die forging, (2) impression die or closed die forging, and (3) flashless forging.

Open die forging involves placing a solid cylindrical piece between two flat (open) dies and reducing its height by compressing it. Closed die or impression die forging involves the use of closed die so that the metal workpiece takes the shape of the die cavity. *Flashless* forging is a special closed die forging in which no excess flash is produced.

Upset forging: *Upsetting*, or *cold heading* is a special forging operation that is commonly used to manufacture bolts.

6.2.2 Mathematical Modeling of Forging

A power (drop) hammer uses air or steam power to accelerate the ram during down-stroke for effective forging operation. The total energy required during down-stroke is given by

$$\text{Total energy} = (mgh) + (pAh) = (mg + pA)h \tag{6.15}$$

where m is mass of ram, kg; g is gravitational acceleration ($=9.81$ m/s^2); p is the air or steam pressure acting on ram cylinder during down-stroke, Pa; A is the cross-sectional area of the ram cylinder [$=(\pi/4)D^2$], m^2; and h is the height of the ram drop, m.

Figure 6.3 presents a simple engineering analysis of an open die forging so as to enable us to mathematically model the forging operation.

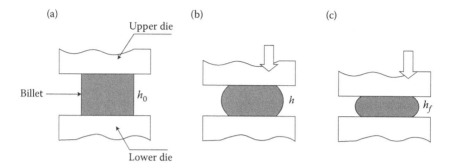

FIGURE 6.3
Changes in geometry of work in an open die forging—(a) start of the forging, (b) forging in process, and (c) final stage of forging.

The *true strain*, ε, in the workpiece during the forging process can be determined by

$$\varepsilon = \ln(h_0/h) \qquad (6.16)$$

where h_0 is the starting height, mm; and h is the height at some point during compression, mm.

The forging force, F_f, can be computed by the following formula:

$$F_f = K_f Y_f A \qquad (6.17)$$

where F_f is the forging force, N; A is the cross-sectional area of the work, mm^2 (the area A continuously increases during the operation as height is reduced); Y_f is the flow stress corresponding to the true strain (ε), MPa; and K_f is the forging shape factor.

The forging shape factor can be computed by the formula

$$K_f = 1 + [(0.4\,\mu D)/h] \qquad (6.18)$$

where μ is the coefficient of friction; D is work-part diameter or other dimension representing contact length with die surface, mm; h is either intermediate or the final height of the work, mm.

In case of *hot forging* (HW), the flow stress (Y_f) is given by

$$Y_f = \text{yield strength} = \sigma_{y.s} \qquad (6.19)$$

In case of *cold forging* (CW), the flow stress (Y_f) is given by

$$Y_f = K\varepsilon^n \qquad (6.20)$$

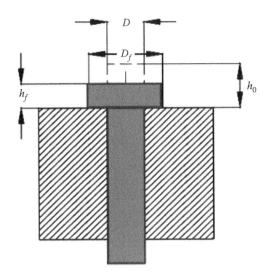

FIGURE 6.4
Geometric design parameters during bolt manufacture.

where K is the strength coefficient, MPa; n is the strain hardening exponent; and ε is true strain.

Figure 6.4 shows the changes in the geometric design parameters of work during bolt manufacture by *upsetting*. The starting material is a bar stock. Since volume of work remains constant, the following relation holds good:

$$(\pi/4)D_0^2 h_0 = (\pi/4)D_f^2 h_f \tag{6.21}$$

where h_0 is the projection length out of the die; h_f is the thickness of the head (of the bolt); D_0 is the original diameter of the bar stock; and D_f is the head diameter.

6.3 Metal Extrusion

6.3.1 Extrusion and Its Types

Extrusion is a metal forming process in which the workpiece is forced to flow through a die opening to produce a long part of desired uniform cross section. Aluminum is the most extensively extruded metal (Saha, 2000). Examples of extruded products include solid hexagonal rods, welding electrodes, and other complex shapes. Depending on the working temperature, extrusion may be either hot or cold extrusion.

Based on deformation, there are two types of extrusion: (1) direct extrusion and (2) indirect extrusion. In *direct extrusion*, a ram compresses the material, forcing it to flow through one or more openings in a die in the same direction as the ram motion. In *indirect extrusion*, the ram penetrates into the work so that metal is forced to flow through the die opening in a direction opposite to ram motion.

6.3.2 Mathematical Modeling/Engineering Analysis of Extrusion

The extrusion process can be mathematically modeled by considering the geometric design features of an extrusion (direct extrusion) machine (see Figure 6.5).

In extrusion, an important process parameter is the *extrusion ratio* that is given by

$$r_x = A_0/A_f \tag{6.22}$$

where r_x is the extrusion ratio; A_0 is the original cross-sectional area of work, mm^2; and A_f is the cross-sectional area of the extruded part.

True strain (ε) is related to *extrusion ratio* by the formula

$$\varepsilon = \ln r_x = \ln (A_0/A_f) \tag{6.23}$$

Another process parameter is the *extrusion strain*, ε_x, which is given by Johnson's formula:

$$\varepsilon_x = a + b\varepsilon = a + b \ln r_x \tag{6.24}$$

where a and b are constants in the Johnson's formula.

The expression for *pressure* depends on whether extrusion is direct or indirect. It also depends whether the die orifice is circular or complex geometry.

In case of *direct extrusion* involving circular die orifice, *ram pressure* is given by

$$p_{d.e.} = \overline{Y}_f \left[\varepsilon_x + (2L/D_0) \right] \tag{6.25}$$

where $p_{d.e.}$ is the ram pressure for direct extrusion, MPa; $\overline{Y}_f$ is the average flow stress, MPa; ε_x is extrusion strain; L is work length, mm; and D_0 is the original diameter of the work, mm (see Figure 6.5).

In case of *indirect extrusion* involving a circular die orifice, *pressure* is given by

$$p_{i.e} = \overline{Y}_f \varepsilon_x \tag{6.26}$$

where $p_{i.e}$ is the pressure for *indirect extrusion*, MPa.

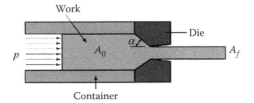

FIGURE 6.5
Geometric design features of extrusion equipment (α = die angle).

In Equations 6.25 and 6.26, the average flow stress can be calculated by

$$\overline{Y}_f = K\varepsilon^n/(1+n) \tag{6.27}$$

where K is the strength coefficient, MPa; ε is true strain; and n is the work-hardening exponent.

In case of *direct extrusion* involving complex die orifice, *ram pressure* is given by

$$p'_{d.e.} = K_s\overline{Y}_f\left[\varepsilon_x + (2L/D_0)\right] \tag{6.28}$$

where K_s is the shape factor for extrusion (Altan et al., 1983), and the other symbols have their usual meanings.

In case of *indirect extrusion* involving complex die orifice, *pressure* is given by

$$p'_{i.e} = K_s\overline{Y}_f\,\varepsilon_x \tag{6.29}$$

Ram force can be computed by the relation

$$F = pA_0 \tag{6.30}$$

where F is the ram force, N; p is the ram pressure, MPa (for direct extrusion, p can be found by using either Equation 6.25 or Equation 6.28; for indirect extrusion, p can be found by using either Equation 6.26 or Equation 6.29); and A_0 is the original cross-sectional area, mm^2 (see Figure 6.5).

Power required for extrusion can be calculated by

$$P = Fv \tag{6.31}$$

where P is the power, W; F is the ram force, N; and v is the ram velocity, m/s.

6.4 Bar/Wire Drawing

Wire drawing is a metal forming operation in which the cross section of a bar or wire is reduced by pulling it through a die opening (Wright, 2010). It is usually done as CW.

Figure 6.6 illustrates the geometric design parameters of a wire/bar drawing equipment so as to enable us to mathematically model the wire drawing operation.

Area reduction, r, in bar drawing is given by

$$r = (A_0 - A_f)/A_0 \tag{6.32}$$

Draft (d, mm) in bar drawing is given by

$$d = D_0 - D_f \tag{6.33}$$

True strain, ε, is given by

$$\varepsilon = \ln (A_0/A_f) \tag{6.34}$$

In case of homogeneous deformation, *draw stress* is given by

$$\sigma = \overline{Y}_f \varepsilon \tag{6.35}$$

where σ is the draw stress, MPa (in case of homogeneous deformation); and $\overline{Y}_f$ is the average flow stress, MPa (given by Equation 6.27).

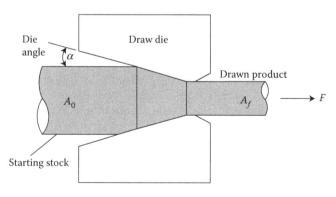

FIGURE 6.6
Geometric design features of wire drawing (L_c = contact length; α = die angle).

Draw force, F, is given by

$$F = \sigma A_f \tag{6.36}$$

where F is the draw force, N; and A_f is the final area (of drawn part), mm^2.
   Actual draw stress in bar drawing is given by the Schey equation as follows:

$$\sigma_d = \phi \varepsilon \overline{Y}_f \left[1 + (\mu / \tan \alpha) \right] \tag{6.37}$$

where σ_d is the actual draw stress, MPa; ϕ is the factor for nonhomogeneous deformation; ε is the *true strain* given by Equation 6.34; μ is the die-work coefficient of friction; and α is the die angle.
   The factor for nonhomogeneous deformation, ϕ, can be determined by

$$\phi = 0.88 + [0.12(D/L_c)] \tag{6.38}$$

where D is the average diameter, mm $[D = (D_0 + D_f)/2]$; and

$$L_c = (D_0 - D_f)/(2 \sin \alpha) \tag{6.39}$$

Once the draw stress, σ_d, has been computed by using Equation 6.37, the draw force, F_d, can be calculated by the formula (see Figure 6.6)

$$F_d = \sigma_d A_f \tag{6.40}$$

6.5 Problems and Solutions—Examples in Bulk Deformation

6.5.1 Examples in Rolling

EXAMPLE 6.1: COMPUTING TRUE STRAIN, DRAFT, AND PERCENTAGE REDUCTION IN ROLLING

A 45 mm thick steel plate was rolled to 38 mm thickness and then again rolled down to 30 mm thickness. Calculate the following: (a) true strain, (b) draft, and (c) reduction in the plate.

Solution

$t_0 = 45$ mm, $t_f = 30$ mm, $\varepsilon = ?$, $d = ?$, reduction $= ?$, % reduction.

(a) By using Equation 6.5,

$$\varepsilon = \ln (t_0/t_f) = \ln (45/30) = 0.4$$

True strain $= 0.4$.

(b) By using Equation 6.2,

$$d = t_0 - t_f = 45\text{--}30 = 15 \, \text{mm}$$

Draft $= 15 \, \text{mm}$.

(c) By reference to Equation 6.3,

$$\text{Reduction} = d/t_o = 15/45 = 0.33$$

Reduction $= 33\%$.

EXAMPLE 6.2: CALCULATING EXIT VELOCITY OF WORK AND FORWARD SLIP

A 38 mm thick plate made of annealed copper is to be reduced to 32 mm thickness in one pass in a rolling operation. As a result of reduction in the thickness, there is a corresponding increase in the width of the plate by 3%. The entrance speed of the plate is 16 m/min. The roll radius is 80 mm, and rotational speed is 30 rev/min. Calculate the following: (a) exit velocity of the plate and (b) forward slip.

Solution

$t_0 = 38 \, \text{mm}, \quad t_f = 32 \, \text{mm}, \quad v_0 = 16 \, \text{m/min} = 16{,}000 \, \text{mm/min}, \quad N = 30 \, \text{rev/min}, \, R = 80 \, \text{mm}, \, v_f = ?, \, S = ?$

$w_f = w_0 + 3\%w_0 = 1.03 \, w_0; \, v_r = R \, \omega_r = R(2\pi N) = 80(2\pi \times 30) = 15{,}080 \, \text{mm/min} = 15.1 \, \text{m/min}$

(a) By using Equation 6.1,

$$t_0 w_0 L_0 = t_f (1.03 w_0) L_f$$

$$t_0 L_0 = t_f (1.03) L_f$$

Dividing both sides by time, t,

$$t_0 v_0 = t_f (1.03) v_f$$

$$38 \times 16{,}000 = 32 \times 1.03 v_f$$

$$608{,}000 = 32.96 v_f$$

$$v_f = 18446.6 \, \text{mm/min} = 18.45 \, \text{m/min}$$

The exit velocity of the plate $= 18.4 \, \text{m/min}$.

(b) By using Equation 6.4,

$$S = (v_f - v_r)/v_r = (18.4\text{--}15.1)/15.1 = 0.21$$

Forward slip $= 0.21$.

EXAMPLE 6.3: CALCULATING THE REQUIRED COEFFICIENT OF FRICTION FOR SUCCESSFUL ROLLING

Using the data in Example 6.2, determine the minimum required coefficient of friction so as to make this rolling operation feasible.

Solution

$t_0 = 38$ mm, $t_f = 32$ mm, $R = 80$ mm, $\mu = ?$

By using Equation 6.2,

$$d = t_0 - t_f = 38-32 = 6 \text{ mm}$$

By reference to inequality (6.11), d_{max} must be greater than d (for feasible rolling).

Taking $d_{max} = 8$ mm, and by using Equation 6.10,

$$d_{max} = \mu^2 R$$

$$8 = \mu^2 (80)$$

$$\mu = 0.32$$

For feasible rolling, the minimum required coefficient of friction $= 0.32$.

EXAMPLE 6.4: DECIDING WHETHER OR NOT ROLLING IS FEASIBLE AND COMPUTING ROLLING FORCE

A 27 mm thick, 280 mm wide metal strip is fed through a two-high rolling mill; each roll has a radius of 240 mm having a rotational speed of 53 rpm. The strip thickness is to be reduced to 24 mm in one pass. The strength coefficient of the work material is 270 MPa and strain hardening exponent is 0.12. The coefficient of friction between the strip and roll surfaces is 0.6.
(a) Is the rolling feasible (successful)?
(b) Determine the rolling force.

Solution

$t_0 = 27$ mm, $t_f = 24$ mm, $w = 280$ mm, $R = 240$ mm, $K = 270$ MPa, $n = 0.12$, $\mu = 0.6$, $F = ?$

(a) By using Equation 6.2,

$$d = t_0 - t_f = 27 - 24 = 3 \text{ mm}$$

By using Equation 6.10,

$$d_{max} = \mu^2 R = (0.6)^2 (240) = 86.4 \text{ mm}$$

Since $d_{max} > d$, rolling is feasible.

(b) By using Equation 6.5,

$$\varepsilon = \ln{(t_0/t_f)} = \ln{(27/24)} = 0.117$$

By using Equation 6.8,

$$\overline{Y}_f = (K\varepsilon^n)/(1+n) = [270(0.117)^{0.12}]/(1+0.12)$$

$$= (270 \times 0.77)/1.12 = 186.5 \, \text{MPa}$$

By using Equation 6.9,

$$F = \overline{Y}_f w(R\,d)^{1/2} = 186.5 \times 280\,(240 \times 3)^{1/2} = 52220 \times 26.8$$

$$= 1401209 \, \text{N} = 1.4 \times 10^6 \, \text{N}$$

Rolling force $= 1.4 \, \text{MN}$.

EXAMPLE 6.5: CALCULATING TORQUE AND POWER FOR ROLLING

By using the data in Example 6.4, calculate the (a) torque and (b) power required for rolling.

Solution

$d = 3 \, \text{mm}$, $w = 280 \, \text{mm}$, $R = 240 \, \text{mm}$, $K = 270 \, \text{MPa}$, $n = 0.12$, $\mu = 0.6$, $F = 1.4 \times 10^6 \, \text{N}$.

(a) By using Equation 6.6

$$L = (Rd)^{1/2} = (240 \times 3)^{1/2} = 26.8 \, \text{mm} = 26.8 \times 10^{-3} \, \text{m}$$

By using Equation 6.13,

Rolling torque $= 2T = FL = 1.4 \times 10^6 \times 26.8 = 37.566 \, \text{MN-mm}$.

Rolling torque $= 37.57 \, \text{kN-m}$.

(b) Roll rotational speed $= N = 53 \, \text{rev/min} = 53/60 \, \text{rev/s} = 0.88 \, \text{rev/s}$.

By using Equation 6.14,

$$\text{Power} = 2\pi NFL = 2\pi \times 0.88 \times 1.4 \times 10^6 \times 26.8 \times 10^{-3}$$
$$= 207.4 \times 10^3 \, \text{W} = 207.4 \, \text{kW}$$

Power required for rolling $= 207.4 \, \text{kW}$.

EXAMPLE 6.6: DETERMINING NUMBER OF ROLLING PASSES REQUIRED AND DRAFT FOR EACH PASS

A reversing two-high mill with each roll's radius of 700 mm is to be used to reduce the thickness of a metal plate from 45 mm to 22 mm by using several passes; the manufacturing process design requires draft to be equal in each pass. The coefficient of friction between rolls and the plate is 0.15. How many rolling passes are required? Also calculate draft for each pass.

Solution

$t_0 = 45$ mm, $t_f = 22$ mm, $R = 700$ mm, $\mu = 0.15$, Drafts are equal, Number of passes = ?

Assuming one pass

$$d = t_0 - t_f = 45 - 22 = 23 \text{ mm}$$

$$d_{max} = \mu^2 R = (0.15)^2 (700) = 15.75 \text{ mm}$$

Since $d_{max} < d$, rolling is *not* feasible with one pass.

Assuming two passes

Let the intermediate thickness be t:

$$\text{Draft for first pass} = d_1 = t_0 - t = 45 - t$$

$$\text{Draft for second pass} = d_2 = t - t_f = t - 22$$

Since draft is to be equal in each pass, $d_1 = d_2$:

$$45 - t = t - 22$$

$$t = 33.5 \text{ mm}$$

Hence, draft for first pass $= d_1 = 45 - t = 45 - 33.5 = 11.5$ mm.

Draft for second pass $= 11.5$.

$d_{max} = 15.7$ mm.

Since $d_{max} > d$, rolling is feasible with two passes.

Hence, it is confirmed that number of rolling passes required $= 2$.

6.5.2 Examples in Forging

EXAMPLE 6.7: CALCULATING TOTAL ENERGY REQUIRED FOR FORGING

A power (drop) forging hammer has a 125 kgf ram. The ram's down-stroke is accelerated by steam pressure of 750 kPa. The ram has a rise of height of 1.5 m. The ram cylinder has a diameter of 10 cm. Calculate the total energy required per blow for forging.

Solution

$m = 125$ kg, $p = 700 \times 10^3$ Pa, $h = 1.5$ m, Ram diameter $= D = 10$ cm $= 0.1$ m, Total Energy = ?

Cross-sectional area of cylinder ram $= A = (\pi/4)D^2 = (\pi/4)\,(0.1)^2 = 7.8 \times 10^{-3}\,m^2$

By using Equation 6.15,

$$\text{Total energy} = (m\,g + p\,A)\,h = [(125 \times 9.81) + (700 \times 10^3 \times 7.8 \times 10^{-3})]\,(1.5)$$
$$= 10{,}029\,J = 10\,kJ$$

Total energy required per blow for forging $= 10\,kJ$.

EXAMPLE 6.8: DETERMINING THE FORGING FORCE AT THE BEGINNING OF THE FORGING OPERATION

A cold upset (open die) forging operation was performed on a 70 mm high cylindrical workpiece having a diameter of 45 mm. The work material has a strength coefficient of 347 MPa, and the strain hardening exponent is 0.16. The coefficient of friction is 0.12.

Determine the forging force as the operation begins (take true strain to be 0.002).

Solution

$h_0 = 70\,mm$, $h = h_0 = 70\,mm$, $D = D_0 = 45\,mm$, $K = 347\,MPa$, $n = 0.16$, $\mu = 0.12$, $\varepsilon = 0.002$.

By using Equation 6.20,

$$Y_f = K\,\varepsilon^n = (347)\,(0.002)^{0.16} = 128.4\,MPa$$

By using Equation 6.18,

$$K_f = 1 + [(0.4\,\mu D)/h] = 1 + \{[(0.4 \times 0.12 \times 45)]/70\} = 1 + 0.031 = 1.03$$

Area at the beginning of the process $= A = A_0 = (\pi/4)D^2 = (\pi/4)\,(45)^2 = 1590.4\,mm^2$.

By using Equation 6.17,

$$F_f = K_f\,Y_f\,A = 1.03 \times 128.4 \times 1590.4 = 210{,}338\,N = 210.3\,kN$$

Forging force as the process begins $= 210.3\,kN$.

EXAMPLE 6.9: DETERMINING FORGING FORCE AT AN INTERMEDIATE STAGE DURING THE FORGING OPERATION

By using the data in Example 6.8, determine the forging force at an intermediate height of 50 mm during the forging operation.

Solution

$h_0 = 70\,mm$, $h = 50\,mm$, $D_0 = 45\,mm$, $K = 347\,MPa$, $n = 0.16$, $\mu = 0.12$.

We need to know the value of D, so we refer to the constant-volume relationship:

$$(\pi/4)D_0^2\, h_0 = (\pi/4)D^2\, h$$

$$(\pi/4)\,(45)^2\,(70) = (\pi/4)\,D^2\,(50)$$

$$D = 53.2\,\text{mm}$$

By using Equation 6.16,

$$\varepsilon = \ln\,(h_0/h) = \ln\,(70/50) = 0.336$$

$$Y_f = K\,\varepsilon^n = (347)\,(0.336)^{0.16} = 291.5\,\text{MPa}$$

$$K_f = 1 + [(0.4\,\mu D)/h] = 1 + \{[(0.4 \times 0.12 \times 53.2)]/50\} = 1 + 0.05 = 1.05$$

Area at the intermediate stage $= A = (\pi/4)D^2 = (\pi/4)\,(53.2)^2 = 2222.8\,\text{mm}^2$.
By using Equation 6.17,

$$F_f = K_f\,Y_f\,A = 1.05 \times 291.5 \times 2222.8 = 681{,}038\,\text{N} = 681\,\text{kN}$$

The forging force at a height of 50 mm during the forging operation $=$ 681 kN.

EXAMPLE 6.10: DETERMINING FORGING FORCE AT THE FINAL STAGE OF THE FORGING OPERATION

By using the data in Example 6.8, determine the forging force at the final height of 32 mm.

Solution

$h_0 = 70\,\text{mm}$, $h = h_f = 32\,\text{mm}$, $D_0 = 45\,\text{mm}$, $K = 347\,\text{MPa}$, $n = 0.16$, $\mu = 0.12$.

We need to know the value of D_f, so we refer to the constant-volume relationship:

$$(\pi/4)D_0^2\, h_0 = (\pi/4)D_f^2\, h_f$$

$$(\pi/4)\,(45)^2\,(70) = (\pi/4)\,D_f^2\,(32)$$

$$D_f = 66.5\,\text{mm}$$

By using Equation 6.16,

$$\varepsilon = \ln\,(h_0/h) = \ln\,(70/32) = 0.78$$

$$Y_f = K\,\varepsilon^n = (347)\,(0.78)^{0.16} = 333.5\,\text{MPa}$$

$$K_f = 1 + [(0.4\,\mu D)/h] = 1 + \{[(0.4 \times 0.12 \times 66.5)]/32\} = 1 + 0.1 = 1.1$$

Area at the final stage $= A_f = (\pi/4)D_f^2 = (\pi/4)\,(66.5)^2 = 3473.2\ \text{mm}^2$.

$$F_f = K_f\,Y_f\,A = 1.1 \times 333.5 \times 3473.2 = 1{,}274{,}051\ \text{N} = 1.27 \times 10^6\ \text{N}$$

The forging force at the final workpiece height $= 1.27$ MN.

EXAMPLE 6.11: EXPLAINING VARIATION OF FORGING FORCE VALUES WITH DEFORMATION

Refer to the answers obtained in Examples 8.8 through 8.10. What is the trend in changes in forging force values with progress of the forging process? Provide a scientific explanation.

Solution

The forging force as the process begins $= 210.3$ kN.

The forging force at a height of 50 mm during the forging operation $=$ 686.8 kN.

The forging force at the final workpiece height $= 1.27$ MN.

It is clearly seen that the forging force increases with the progress of the forging process. This increasing trend in the forging force is due to strain hardening because the process is cold forging/CW.

EXAMPLE 6.12: CALCULATING BOLT-HEAD THICKNESS AND FORGING SHAPE FACTOR

A cold heading forging operation is performed to produce the head on a steel bar. The strength coefficient of the material is 480 MPa, and the strain hardening exponent is 0.19. Coefficient of friction at the die–work interface is 0.11. The diameter of bar stock is 5.5 mm, and its projection length out of the die is 4.5 mm. The head is to have a diameter of 7 mm. Calculate the following: (a) the thickness of the head and (b) forging shape factor.

Solution

By reference to Figure 6.6, $h_0 = 4.5$ mm, $h_f = ?$, $D_0 = 5.5$ mm, $D_f = 7$ mm, $\mu = 0.11$.

By using Equation 6.21,

$$(\pi/4)D_0^2\,h_0 = (\pi/4)D_f^2\,h_f$$

$$(\pi/4)\,(5.5)^2\,(4.5) = (\pi/4)\,(7)^2\,h_f$$

$$h_f = 2.78$$

Thickness of the head $= h_f = 2.78$ mm.

By using Equation 6.18,

$$K_f = 1 + [(0.4\,\mu D)/h_f] = 1 + [(0.4 \times 0.11 \times 5.5)/2.78] = 1 + 0.08 = 1.08$$

Forging shape factor $= K_f = 1.08$.

EXAMPLE 6.13: CALCULATING THE PUNCH FORCE FOR BOLT FORGING

By using the data in Example 6.12, calculate the following: (a) true strain, (b) flow stress, and (c) the maximum punch force to form the head.

Solution

$K = 480\,\text{MPa}$, $n = 0.19$, $h_0 = 4.5\,\text{mm}$, $h_f = 2.78\,\text{mm}$, $K_f = 1.08$, $D_f = 7\,\text{mm}$.

(a) By using Equation 6.16,

$$\varepsilon = \ln\,(h_0/h) = \ln\,(4.5/2.78) = 0.48$$

True strain $= \varepsilon = 0.48$.

(b) By using Equation 6.20,

$$Y_f = K\,\varepsilon^n = (480)\,(0.48)^{0.19} = 417.8\,\text{MPa}$$

Flow stress $= Y_f = 417.8\,\text{MPa}$.

(c) In order to form the head, we must consider area of the head (i.e., diameter of head, D_f):

$$\text{Area of head} = A = (\pi/4)D_f^2 = (\pi/4)\,(7)^2 = 38.48\,\text{mm}^2$$

By using Equation 6.17,

$$F_f = K_f\,Y_f\,A = 1.08 \times 417.8 \times 38.48 = 17{,}365.1\,\text{N}$$

The maximum punch force to form the head $= 17{,}365\,\text{N}$.

EXAMPLE 6.14: CALCULATING THE PUNCH FORCE IN HOT FORGING

A hot upset (open die) forging operation is performed on a metal workpiece having initial diameter of 30 mm and initial height of 55 mm. The workpiece is upset to a final diameter $= 52\,\text{mm}$. The yield strength of the work metal at the elevated temperature is 87 MPa. The coefficient of friction between the die and work surfaces is 0.38. Calculate the maximum punching force for the forging operation.

Solution

$h_0 = 55\,\text{mm}$, $h_f = ?$, $D_0 = 30\,\text{mm}$, $D_f = 52\,\text{mm}$, $S_y = 87\,\text{MPa}$, $\mu = 0.38$, $F_f = ?$

In order to calculate the forging force, we need to know the flow stress (Y_f), the final work area (A_f), and forging shape factor (K_f); the latter requires determination of the final height, h_f.

By using Equation 6.21,

$$(\pi/4)D^2\, h_0 = (\pi/4)D_f^2\, h_f$$

$$(\pi/4)\, (30)^2\, (55) = (\pi/4)\, (52)^2\, h_f$$

$$h_f = 18.3\,\text{mm}$$

For hot forging, we must use Equation 6.19:

$$Y_f = \text{yield strength} = S_y = 87\,\text{MPa}$$

By using Equation 6.18,

$$K_f = 1 + [(0.4\,\mu D_f)/h_f] = 1 + [(0.4 \times 0.38 \times 52)/18.3] = 1.43$$

The maximum forging force corresponds to area, A_f [$A_f = (\pi/4)D_f^2 = (\pi/4)\,(52)^2 = 2123.7\,\text{mm}^2$].

By using Equation 6.17,

$$F_f = K_f\, Y_f\, A = 1.43 \times 87 \times 2123.7 = 264{,}211.5\,\text{N} = 264.2\,\text{kN}$$

The maximum punching force for the forging operation $= 264.2\,\text{kN}$.

EXAMPLE 6.15: CALCULATING THE MAXIMUM FORGING FORCE FOR A METAL STRIP

An open die forging operation is performed on a lead strip with initial dimensions 24 mm × 24 mm × 150 mm at room temperature (25°C). The final strip dimensions are 6 mm × 96 mm × 150 mm. The yield strength of lead is 7 MPa, and the coefficient of friction at the die–work interface is 0.25. Compute the maximum forging force.

Solution

$h_0 = 24\,\text{mm}$, $w_0 = 24\,\text{mm}$, $L_0 = 150\,\text{mm}$; $h_f = 6\,\text{mm}$, $w_f = 96\,\text{mm}$, $L_f = 150\,\text{mm}$, $S_y = 7\,\text{MPa}$, $\mu = 0.25$.

Since forging of lead at room temperature is actually HW,

$$Y_f = \text{yield strength} = S_y = 7\,\text{MPa}$$

The maximum forging force corresponds to the final area:

$$\text{Final area} = A = 96 \times 150 = 14{,}400\,\text{mm}^2$$

$$D^2 = 96^2 + 150^2,\ D = 178\,\text{mm}$$

$$K_f = 1 + [(0.4\,\mu D)/h] = 1 + \{[(0.4 \times 0.25 \times 178)]/6\} = 1 + 2.96 = 3.96$$

By using Equation 6.17,

$$F_f = K_f \, Y_f \, A = 3.96 \times 7 \times 14{,}400 = 399{,}168 \, \text{N} = 0.4 \times 10^6 \, \text{N}$$

The maximum forging force $= 400$ kN.

6.5.3 Examples in Extrusion

EXAMPLE 6.16: CALCULATING EXTRUSION RATIO, TRUE STRAIN, AND EXTRUSION STRAIN

A cylindrical billet that is 90 mm long and 55 mm in diameter is reduced by direct extrusion to a 25 mm diameter. Determine the following: (a) extrusion ratio, (b) true strain, and (c) extrusion strain (the Johnson equation has $a = 0.7$ and $b = 1.5$).

Solution

$D_0 = 55$ mm, $D_f = 25$ mm, Johnson's formula: $a = 0.7$, $b = 1.5$, $r_x = ?$, $\varepsilon = ?$, $\varepsilon_x = ?$

(a) The cross-sectional area before and after extrusion can be found as follows:

$$A_0 = (\pi/4) \, D_0^2 = (\pi/4) \, (55)^2 = 2375.8 \, \text{mm}^2$$

$$A_f = (\pi/4) \, D_0^2 = (\pi/4) \, (25)^2 = 490.8 \, \text{mm}^2$$

By using Equation 6.22,

$$r_x = A_0/A_f = (2375.8)/(490.8) = 4.8$$

Extrusion ratio $= r_x = 4.8$.

(b) By using Equation 6.23,

$$\varepsilon = \ln r_x = \ln 4.8 = 1.577$$

True strain $= \varepsilon = 1.577$.

(c) By using Johnson's formula, Equation 6.24,

$$\varepsilon_x = a + b \, \varepsilon = 0.7 + (1.5 \times 1.577) = 3$$

Extrusion strain $= \varepsilon_x = 3$.

EXAMPLE 6.17: CALCULATING RAM PRESSURE FOR DIRECT EXTRUSION

By using the data in Example 6.16, calculate the *ram pressure* if the cylindrical billet's material is annealed mild steel. Hint: Refer to the data in Table 5.1.

Solution

By reference to Example 6.16, the direct extrusion data: $D_0 = 55$ mm, $L = 90$ mm, $\varepsilon = 1.577$, $\varepsilon_x = 3$.

By reference to the data for annealed mild steel in Table 5.1, we get $n = 0.22$, $K = 600$ MPa.

By using Equation 6.27,

$$\overline{Y}_f = K\varepsilon^n/(1+n) = [600(1.58)^{0.22}]/(1+0.22) = (600 \times 1.1)/1.22$$
$$= 543.8 \text{ MPa}$$

By using Equation 6.25 (for direct extrusion),

$$p_{d.e.} = \overline{Y}_f [\varepsilon_x + (2L/D_0)] = (543.8)\{3 + [(2 \times 90)/55]\}$$
$$= (543.8)(3 + 3.27) = 3411.1 \text{ MPa}$$

Ram pressure $= p_{d.e.} = 3411.1$ MPa.

EXAMPLE 6.18: CALCULATING THE RAM PRESSURE FOR INDIRECT EXTRUSION

An annealed 0.05%C steel cylindrical billet that is 90 mm long and 55 mm in diameter is reduced by indirect extrusion to a 25 mm diameter. The Johnson equation has $a = 0.7$ and $b = 1.5$.

Determine the *ram pressure*.

Solution

$D_0 = 55$ mm, $D_f = 25$ mm. From Table 5.1: $n = 0.22$, $K = 600$ MPa, $p_{i.e} = ?$

$$A_0 = (\pi/4) D_0^2 = (\pi/4)(55)^2 = 2375.8 \text{ mm}^2$$

$$A_f = (\pi/4) D_0^2 = (\pi/4)(25)^2 = 490.8 \text{ mm}^2$$

$$r_x = A_0/A_f = (2375.8)/(490.8) = 4.8$$

$$\varepsilon = \ln r_x = \ln 4.8 = 1.577$$

By using Johnson's formula, Equation 6.24,

$$\varepsilon_x = a + b\,\varepsilon = 0.7 + (1.5 \times 1.577) = 3$$

By using Equation 6.27,

$$\overline{Y}_f = K\varepsilon^n/(1+n) = [600(1.58)^{0.22}]/(1+0.22) = (600 \times 1.1)/1.22 = 543.8 \text{ MPa}$$

Since this is *indirect extrusion* involving *circular die orifice*, we can use Equation 6.26:

$$p_{i.e} = \overline{Y}_f \, \varepsilon_x = 543.8 \times 3 = 1631.4 \, \text{MPa}$$

The ram pressure = 1631.4 MPa.

EXAMPLE 6.19: CALCULATING RAM FORCE AND POWER FOR EXTRUSION

By using the data in Example 6.18, calculate the ram force and power to carry out extrusion if it takes 7 seconds to perform the extrusion operation.

Solution

$p = 1631.4$ MPa, $A_0 = 2375.8 \, \text{mm}^2$, $L = 90 \, \text{mm} = 0.090 \, \text{m}$, time = 40 s.

By using Equation 6.30,

$$F = p \, A_0 = 1631.4 \times 2375.8 = 3{,}875{,}880 \, \text{N}$$

Ram force = $F = 3{,}875{,}880$ N.

Ram velocity = $v = L/t = 0.09/7 = 0.0128$ m/s.

By using Equation 6.31,

$$P = F \, v = 3{,}875{,}880 \times 0.0128 = 49{,}832.7 \, \text{W}$$

Power = 49.8 kW.

EXAMPLE 6.20: COMPUTING RAM PRESSURE FOR EXTRUSION INVOLVING COMPLEX DIE ORIFICE

An annealed 0.05%C steel cylindrical billet that is 90 mm long and 55 mm in diameter is reduced by indirect extrusion to a complex cross section approximating to 25 mm diameter. The shape factor for the extrusion is 1.2. The Johnson equation has $a = 0.7$ and $b = 1.5$. Determine the *ram pressure*.

Solution

$D_0 = 55$ mm, $D_f = 25$ mm. From Table 5.1: $n = 0.22$, $K = 600$ MPa, $K_s = 1.2$, $p_{i.e} = ?$

$$A_0 = (\pi/4) \, D_0^2 = (\pi/4) \, (55)^2 = 2375.8 \, \text{mm}^2$$

$$A_f = (\pi/4) \, D_0^2 = (\pi/4) \, (25)^2 = 490.8 \, \text{mm}^2$$

$$r_x = A_0/A_f = (2375.8)/(490.8) = 4.8$$

$$\varepsilon = \ln r_x = \ln 4.8 = 1.577$$

$$\varepsilon_x = a + b\,\varepsilon = 0.7 + (1.5 \times 1.577) = 3$$

$$\overline{Y}_f = K\varepsilon^n/(1+n) = [600\,(1.58)^{0.22}]/(1+0.22) = (600 \times 1.1)/1.22$$
$$= 543.8\,\text{MPa}$$

Since this is the case of indirect extrusion involving a complex die orifice, we use Equation 6.29:

$$p'_{i.e} = K_s\overline{Y}_f\,\varepsilon_x = 1.2 \times 543.8 \times 3 = 1957.7\,\text{MPa}$$

Ram pressure = 1957.7 MPa.

6.5.4 Examples in Bar/Wire Drawing

EXAMPLE 6.21: CALCULATING AREA REDUCTION, DRAFT, AND TRUE STRAIN IN BAR DRAWING

A bar, with starting diameter of 2.7 mm, is drawn through a draw die to reduce to the final diameter of 2 mm. Calculate the following: (a) area reduction, (b) draft, and (b) true strain.

Solution

$D_0 = 2.7$ mm, $D_f = 2$ mm, $r = ?$, $d = ?$, $\varepsilon = ?$

$$A_0 = (\pi/4)\,D_0^2 = (\pi/4)\,(2.7)^2 = 5.72\,\text{mm}^2$$

$$A_f = (\pi/4)\,D_f^2 = (\pi/4)\,(2)^2 = 3.142\,\text{mm}^2$$

(a) By using Equation 6.32,

$$r = (A_0 - A_f)/A_0 = (5.72 - 3.142)/5.72 = 0.45 = 45\%$$

Area reduction = 45%.

(b) By using Equation 6.33,

$$d = D_0 - D_f = 2.7 - 2 = 0.7$$

Draft = 0.7 mm.

(c) By using Equation 6.34,

$$\varepsilon = \ln(A_0/A_f) = \ln(5.72/3.142) = 0.6$$

True strain = $\varepsilon = 0.6$.

EXAMPLE 6.22: CALCULATING DRAW STRESS AND DRAW FORCE IN IDEAL BAR DRAWING

A bar with starting diameter of 2.7 mm is drawn through a draw die involving homogeneous deformation (ideal bar drawing). The final diameter is 2 mm. The metal has a strength coefficient $K = 210$ MPa and a strain hardening exponent $n = 0.18$. Calculate draw stress and draw force.

Solution

$A_0 = 5.72$ mm^2, $A_f = 3.142$ mm^2, $\varepsilon = 0.6$, $K = 210$ MPa, $n = 0.18$, $\sigma = ?$, $F = ?$

By using Equation 6.27,

$$\overline{Y}_f = K\varepsilon^n/(1+n) = [210\,(0.6)^{0.18}]/(1+0.18) = 191.5/1.18 = 162.28 \text{ MPa}$$

By using Equation 6.35 (case of homogeneous deformation),

$$\sigma = \overline{Y}_f\,\varepsilon = 162.28 \times 0.6 = 97.37 \text{ MPa}$$

Draw stress $= 97.4$ MPa.

By using Equation 6.40,

$$F_d = \sigma_d\,A_f = 97.37 \times 3.142 = 305.89 \text{ N}$$

Draw force $= 305.9$ N.

EXAMPLE 6.23: CALCULATING DRAW STRESS AND DRAW FORCE IN NONIDEAL WIRE DRAWING

A wire is drawn through a draw die with an entrance angle (die angle) of 20°. The deformation is nonhomogeneous. Starting diameter of the bar is 2.6 mm, and the final diameter is 1.8 mm. The coefficient of friction at the work–die interface is 0.06. The metal has a strength coefficient $K = 208$ MPa and a strain hardening exponent $n = 0.20$. Determine the draw stress and draw force.

Solution

$D_0 = 2.6$ mm, $D_f = 1.8$ mm, $\alpha = 20°$, $\mu = 0.06$, $\sigma_d = ?$, $F = ?$

The average diameter $= D = (D_0 + D_f)\,/2 = (2.6 + 1.8)/2 = 2.2$ mm:

$$A_0 = (\pi/4)\,D_0^2 = (\pi/4)\,(2.6)^2 = 5.31 \text{ mm}^2$$

$$A_f = (\pi/4)\,D_f^2 = (\pi/4)\,(1.8)^2 = 2.54 \text{ mm}^2$$

By using Equation 6.34,

$$\varepsilon = \ln\,(A_0/A_f) = \ln\,(5.31/2.54) = 0.73$$

By using Equation 6.39,

$$L_c = (D_0 - D_f)/(2 \sin \alpha) = (2.6-1.8)/(2 \sin 20°) = (0.8)/(0.68) = 1.17$$

The factor for nonhomogeneous deformation, ϕ, can be determined by using Equation 6.38 as follows:

$$\phi = 0.88 + [0.12 \, (D/L_c)] = 0.88 + [0.12 \, (2.2/1.17)] = 0.88 + 0.225 = 1.1$$

By using Equation 6.27,

$$\bar{Y}_f = K\varepsilon^n/(1+n) = [208(0.73)^{0.20}]/(1+0.20) = 195.7/1.20 = 163.1$$

Actual draw stress, given by the Schey formula (Equation 6.37) can be calculated as follows:

$$\sigma_d = \phi e \bar{Y}_f \, [1 + (\mu/\tan \alpha)] = (1.1 \times 0.73 \times 163.1)[1 + (0.06/\tan 20°)]$$
$$= 152.5 \, \text{MPa}$$

By using Equation 6.40,

$$F_d = \sigma_d \, A_f = 152.5 \times 2.54 = 387.3 \, \text{N}$$

Draw stress = 152.5 MPa.

Draw force = 387.3 N.

Questions and Problems

6.1 Define the following terms: (a) rolling, (b) forging, (c) extrusion, and (d) wire drawing.

6.2 (a) Arrange the following types of rolling mills in the order of decreasing roll radii: three-high mill, cluster mill, two-high mill, four-high mill. (b) Which of the rolling mills in (a) is the most energy efficient? Justify your answer.

6.3 (a) Obtain a mathematical formula for torque for rolling with the aid of a sketch. (b) Derive a mathematical model for power required for rolling.

6.4 Sketch a power drop hammer for forging.

6.5 Which process is the most suitable for manufacturing metal bolts? Describe the process.

6.6 Differentiate between direct extrusion and indirect extrusion with the aid of sketches.

6.7 Which bulk deformation process is carried out as CW? Explain the process with the aid of a sketch.

P6.8 A 50 mm thick steel plate was rolled to 42 mm thickness and then again rolled down to 35 mm thickness. Calculate the following: (a) draft, (b) true strain, and (c) reduction in the plate.

P6.9 A 35 mm thick plate made of annealed copper is to be reduced to 27 mm thickness in one pass in a rolling operation. As a result of reduction in the thickness, there is a corresponding increase in width of the plate by 2%. The entrance speed of the plate is 18 m/min. The roll radius is 85 mm, and the rotational speed is 30 rev/min. Calculate the (a) exit velocity of the plate and (b) forward slip.

P6.10 By using the data in P6.9, determine the minimum required coefficient of friction so as to make this rolling operation feasible.

P6.11 A 26 mm thick 310 mm wide metal strip is fed through a two-high rolling mill; each roll has a radius of 260 mm having a rotational speed of 52 rpm. The strip thickness is to be reduced to 22 mm in one pass. The strength coefficient and strain hardening exponent of the material values are 275 MPa and 0.12, respectively. The coefficient of friction between the strip and roll surfaces is 0.55. (a) Is the rolling feasible (successful)? (b) Determine the rolling force.

P6.12 By using the data in P6.11, calculate the (a) torque and (b) power required for rolling.

P6.13 A reversing two-high mill with each roll's radius of 650 mm is to be used to reduce thickness of a metal plate from 48 mm to 26 mm by using several passes; the manufacturing process design requires draft to be equal in each pass. The coefficient of friction between rolls and the plate is 0.21. How many rolling passes are required? Also calculate draft for each pass.

P6.14 A power (drop) forging hammer has a 120 kgf ram. The ram's downstroke is accelerated by steam pressure of 760 kPa. The ram has a rise of height of 1.3 m. The ram cylinder has a diameter of 12 cm. Calculate the total energy required per blow for forging.

P6.15 A cold upset (open die) forging operation was performed on a 75 mm high cylindrical workpiece having a diameter of 50 mm. The work material has a strength coefficient of 355 MPa, and strain hardening exponent is 0.17. The coefficient of friction is 0.13.
Determine the forging force as the operation begins (take true strain to be 0.002).

P6.16 By using the data in P6.15, determine the forging force at an intermediate height of 57 mm during the forging operation.

P6.17   By using the data in P6.15, determine the forging force at a final height of 40 mm.

P6.18   A cold heading forging operation is performed to produce the head on a steel bar for bolt manufacturing. The strength coefficient and the strain hardening exponent values of the material are 460 MPa and 0.18, respectively. Coefficient of friction at the die–work interface is 0.13. The diameter of bar stock is 6.5 mm, and its projection length out of the die is 5.0 mm. The head is to have a diameter of 8.5 mm. Calculate the following: (a) the thickness of the head and (b) the forging shape factor.

P6.19   A hot upset (open die) forging operation is performed on a metal workpiece having initial diameter of 35 mm and initial height of 52 mm. The workpiece is upset to a final diameter = 48 mm. The yield strength of the work metal at the elevated temperature is 92 MPa. The coefficient of friction between the die and work surfaces is 0.36. Calculate the maximum punching force for the forging operation.

P6.20   An open-die hot forging operation is performed on a metal strip with initial dimensions 25 mm × 25 mm × 150 mm. The final strip dimensions are around 7 mm × 90 mm × 150 mm. The yield strength of the metal at the elevated temperature is 150 MPa, and the coefficient of friction at the die–work interface is 0.23. Compute the maximum forging force.

P6.21   A 95 mm long cylindrical billet having a diameter of 58 mm is reduced to a 26 mm diameter by direct extrusion. Determine the following: (a) extrusion ratio, (b) true strain, and (c) extrusion strain (the Johnson equation has $a = 0.6$ and $b = 1.2$).

P6.22   By using the data in P6.21, calculate the *ram pressure* if the cylindrical billet's material is annealed copper. Hint: Refer to the data in Table 5.1.

P6.23   An annealed brass cylindrical billet that is 85 mm long and 53 mm in diameter is reduced by indirect extrusion to a 27 mm diameter. The Johnson equation has $a = 0.8$ and $b = 1.4$.
Determine the *ram pressure*.

P6.24   By using the data in P6.23, calculate the ram force and power to carry out extrusion if it takes 6 seconds to perform the extrusion operation.

P6.25   An annealed copper cylindrical billet that is 88 mm long and 51 mm in diameter is reduced by indirect extrusion to a complex cross section approximating to 22 mm diameter. The shape factor for the extrusion is 1.15. The Johnson equation has $a = 0.8$ and $b = 1.4$.
Determine the *ram pressure*.

P6.26   A metal bar having a diameter of 2.4 mm is drawn through a draw die to reduce to the final diameter of 1.8 mm. Calculate the following: (a) area reduction, (b) draft, and (b) true strain.

P6.27 A bar with starting diameter of 2.8 mm is drawn through a draw die involving homogeneous deformation (ideal bar drawing). The final diameter is 2.2 mm. The metal has a strain hardening exponent of 0.22, and strength coefficient, $K = 600$ MPa. Calculate (a) draw stress and (b) draw force.

P6.28 A wire is drawn through a draw die with an entrance angle (die angle) of 19°. The deformation is nonhomogeneous. The starting diameter of the bar is 2.3 mm, and the final diameter is 1.4 mm. The coefficient of friction at the work–die interface is 0.06. The metal has a strength coefficient $K = 480$ MPa and a strain hardening exponent $n = 0.19$. Determine the (a) draw stress and (b) draw force.

References

Altan, T., Oh, S-I., Gegel, H. L. 1983. *Metal Forming: Fundamentals and Applications.* Materials Park, OH: ASM International.

Ares, J.A. 2006. *Metal: Forming, Forging, and Soldering Techniques.* New York: Barron's Educational Series.

Ginzburg, V.B. 1989. *Steel-Rolling Technology: Theory and Practice.* Boca Raton, FL: CRC Press.

Ginzburg, V.B., Ballas, R. 2000. *Flat Rolling Fundamentals.* Boca Raton, FL: CRC Press/ Taylor and Francis Group.

Huda, Z. 2017. *Materials Processing for Engineering Manufacture.* Switzerland: Trans Tech.

Lu, C. 2007. Measurement of the forward slip in cold strip rolling using a high speed digital camera. *Journal of Mechanical Science and Technology,* 21(10), 1528–1533.

Saha, P.K. 2000. *Aluminum Extrusion Technology.* Materials Park, OH: ASM International.

Wright, R.N. 2010. *Wire Technology: Process Engineering and Metallurgy.* Oxford, UK: Butterworth-Heinemann.

7

Metal Forming III—Sheet Metal Forming

7.1 Metal Sheets and Plates

A distinction can be made between metal sheet and plate as follows. Sheet metal thicknesses are typically in the range of 0.4–6 mm. When thickness exceeds about 6 mm, the stock is usually referred to as *plate*. Metal sheets and plates are the backbones of automotive and aerospace structures. They are extensively used in the manufacture of bodies of automobiles (including buses, trucks, etc.) and railway cars. They are also used in aircraft structures, farm and construction equipment, small and large tools, and office furniture.

7.2 Types of Sheet Metal Forming Processes

All sheet metal forming processes can be classified into two groups: (1) press-work sheet metal forming processes and (2) non-press-work sheet metal forming operations (e.g., spinning, stretch forming, etc.). The press-work sheet metal forming operations are more commonly practiced and include shearing (open shearing), blanking, piercing, bending, deep drawing, roll bending, and roll forming. These operations are discussed in the following sections.

7.3 Shearing and Related Operations

7.3.1 Distinction between Shearing and Blanking/Piercing

There are three types of shearing operations in sheet metal forming: (a) *open shearing*, (b) *blanking*, and (c) *piercing*. *Shearing*, or *open shearing*, involves cutting off a larger metal sheet into smaller ones (Figure 7.1) (see Section 7.3.2). In a *blanking* operation, the cut-out part is used for further press-work operations. If the cut-out part is a scrap, the operation is called *piercing*.

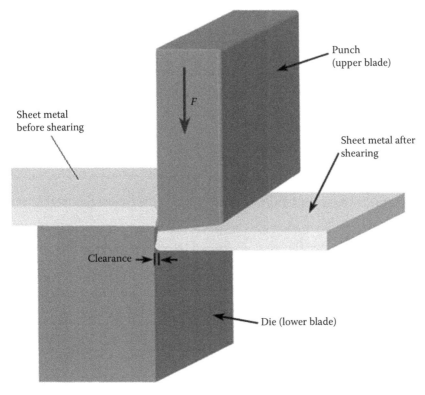

Punch
(upper blade)

Sheet metal
before shearing

Sheet metal after
shearing

F

Clearance

Die (lower blade)

FIGURE 7.1
Shearing operation showing *clearance*.

7.3.2 Shearing and Its Mathematical Modeling

General: Shearing, or *open shearing*, is a sheet metal cut-off operation along a straight line between two cutting edges. *Shearing* is used to cut large sheets into smaller sections. In *shearing*, a piece of sheet metal is separated by applying a *shearing force* that is high enough to cause the material to fail; the failure occurs when the shear stress in the material exceeds the ultimate shear strength of the material (Altan and Tekkaya, 2012). The shearing force is applied by two tools, one above and one below the sheet: the upper tool is called *punch* and the lower tool is called *die* (see Figure 7.1).

Mathematical Modeling: It is important to allow adequate *clearance* in shearing and related operations (see Figure 7.1). The *clearance* depends on the sheet thickness and the type of metal being formed; it is determined by (Huda, 2017)

$$c = A_c t \qquad (7.1)$$

TABLE 7.1

Clearance Allowance (A_c) Data for Shearing of Various Materials

Material	Stainless Steel (half hard and full hard), CR Steel (half hard)	2024ST and 6061ST *Al* Alloys; Brass, Soft CR Steel, Soft Stainless Steel	1100S and 5052S *Al* Alloys, All Tempers
Clearance allowance	0.075	0.06	0.045

where c is the clearance, mm; A_c is the clearance allowance; and t is the thickness of the sheet metal, mm.

The *clearance allowance* (A_c) depends on the type of metal to be sheared; the values of A_c for various groups of metals are given in Table 7.1.

Shearing force or *punching force* depends on the dimensions and strength of the sheet metal; it can be determined by

$$F = \tau \cdot t \cdot L = 0.7\,S_{ut}t \cdot L \qquad (7.2)$$

where F is the punching force, N; t is thickness of the sheet metal, mm; L is the length of cut, mm; τ is the shear strength, MPa; and S_{ut} is the ultimate tensile strength, MPa.

7.3.3 Blanking/Piercing and Their Mathematical Modeling

Blanking is cutting out a large sheet of stock into smaller pieces (blanks) suitable for further press-work operations (see Figure 7.2). *Piercing* or *punching* involves cutting internal features (holes or slots) in stock. In *piercing*, the separated piece is scrap; the remaining stock is the desired part.

Figure 7.2 illustrates the geometric parameters (clearance, c, and punch radius, r) in a blanking/piercing operation. The clearance (c) and the punching force (F) can be determined by using Equations 7.1 and 7.2, respectively.

In a blanking operation, the punch size and die size are determined on the basis of blank size, as follows:

Diameter of a blank $= D_b$.

$$\text{Blanking punch diameter} = D_b - 2c \qquad (7.3)$$

$$\text{Blanking die diameter} = D_b \qquad (7.4)$$

In a punching operation, the punch size and die size are determined on the basis of hole diameter, as follows:

Hole diameter $= D_h$.

$$\text{Piercing punch diameter} = D_h \qquad (7.5)$$

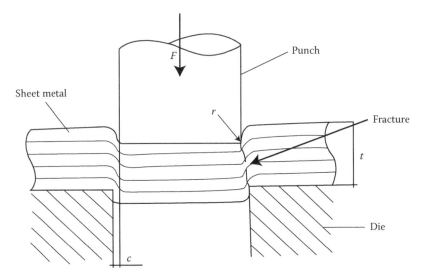

FIGURE 7.2
Blanking/piercing operation showing the geometric design parameters (F = punching force, c = clearance, r = punch radius, t = thickness).

$$\text{Piercing die diameter} = D_h + 2c \qquad (7.6)$$

Compound Die and Progressive Die: A *compound die* enables us to perform several operations (e.g., blanking, piercing, etc.) on the same strip in one stroke at one station. Progressive dies are made with two or more stations; each station performs an operation on the workpiece so that the workpiece is completed when the last operation has been accomplished.

7.4 Bending

7.4.1 Bending and Its Types

Bending is a sheet metal forming process that involves the application of a force to a piece of sheet metal, thereby causing it to bend at an angle and form the desired shape. Bent parts can be quite small, such as a bracket, or very long (up to 20 feet), such as a large chassis. Bending essentially involves straining of the metal sheet around a straight (neutral) axis (see Figure 7.3).

There are two main types of bending operations: (a) *V*-bending and (b) edge bending. In *V-bending*, a *V*-shaped punch forces the work into the *V*-shaped die and, hence, bends it. In edge bending, the punch applies force to a cantilever beam section, causing the work to bend over the edge of the die.

7.4.2 Mathematical Modeling of Bending

Spring-back: When the bending force is removed at the end of deformation, there is an elastic recovery called *spring-back* (*SB*). Numerically, *SB* is defined as the increase in included angle of the bent part relative to the included angle of the punch after the tool (punch) is removed. *SB* is given by

$$SB = (\alpha' - \alpha'_p)/\alpha'_p \qquad (7.7)$$

where α' is the included angle of the sheet metal part, degree; and α'_p is the included angle of the punch, degree. It is desirable to limit *SB* to a low value. Recent research has shown that special-grades titanium and aluminum alloy sheets can be bent to high angles, with very low fillet radii and limited *SB* (Gisario et al., 2016).

Bending Force: The *bending force* is the force required to perform bending; it depends on the strength, thickness, and length of the sheet metal as well as on the geometric design parameters of punch and die. The *bending force* is given by

$$F = (K_{bf}S_{ut}W\,t^2)/D \qquad (7.8)$$

where F is the bending force, N; S_{ut} is the tensile strength of sheet metal, MPa; W is the part width in the direction of the bend axis, mm; t is the stock thickness; D is the die opening dimension, mm; and K_{bf} is the bending factor constant that accounts for the difference between ideal and actual bending processes. The bending factor constant values are given in Table 7.2.

Bend Allowance: A look at Figure 7.3 indicates that the outside portion of the bent sheet metal is under tension (i.e., there is stretching at the outside portion). Hence, it is important to allow an adequate bend allowance, which is given by (Groover, 2014)

$$A_b = [2\pi\alpha(R + K_{ba}t]/360 \qquad (7.9)$$

where A_b is the bend allowance, mm; α is the bend angle ($= 180 - \alpha'$), degrees; t is the stock thickness, mm; R is the bend radius, mm; and K_{ba} is the factor to account for stretching. The value of K_{ba} depends on bend geometry. If $R < 2t$, $K_{ba} = 0.33$; if $R \geq 2t$, $K_{ba} = 0.50$.

TABLE 7.2

Bending Factor Constants for Two Types of Bending

Bending Type	V-Bending	Edge Bending
K_{bf}	1.33	0.33

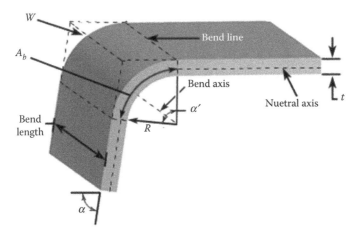

FIGURE 7.3
Bending of sheet metal (Note: $\alpha = 180 - \alpha'$; $W =$ width, $t =$ thickness, $R =$ bend radius, $\alpha =$ bend angle, $\alpha' =$ included angle).

In view of the bend allowance, the blank size for bending can be calculated by

$$L_{blank} = L + A_b \tag{7.10}$$

where L_{blank} is the blank size or length of the blank; L is the total length; and A_b is the bend allowance.

7.5 Deep Drawing

7.5.1 General

Deep drawing is a sheet metal forming operation that involves placing a piece of sheet metal over a die cavity and then pushing the metal into the opening by using a punch. Commonly used deep-drawn products include beverage cans, cooking pots, and other cup-shaped products.

7.5.2 Mathematical Modeling of Deep Drawing

Figure 7.4 illustrates a deep drawing (*DD*) operation with particular reference to its geometric design features.

The clearance depends on the sheet metal (blank) thickness (see Figure 7.4) and is given by

$$c = 1.1t \tag{7.11}$$

where c is the clearance, mm; and t is the blank thickness, mm.

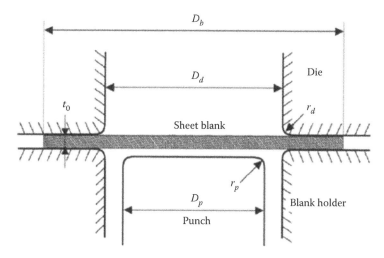

FIGURE 7.4
DD operation (D_p = punch diameter; D_d = die diameter; r_p = punch corner radius; r_d = die corner radius; D_b = blank diameter, c = clearance).

The ratio of the blank diameter to the punch diameter is called the *drawing ratio (DR)*; it measures the severity of a *DD* operation. *DR* is given by

$$DR = D_b/D_p \tag{7.12}$$

where *DR* is the drawing ratio; D_b is the blank diameter, mm; and D_p is the punch diameter, mm.

The *reduction, r,* in *DD* is given by

$$r = (D_b - D_p)/D_b \tag{7.13}$$

The *thickness-to-diameter ratio* in *DD* is given by

$$\text{thickness-to-diameter ratio} = t/D_b \tag{7.14}$$

Feasibility of Deep Drawing: In order for the *DD* operation to be feasible (successful), certain operating conditions must be satisfied; these conditions are given in Table 7.3.

Deep Drawing Force: The force required to deep draw can be calculated by

$$F_{dd} = \pi D_p \cdot t \cdot S_{ut}[(D_b/D_p) - 0.7] \tag{7.15}$$

where F_{dd} is the *DD* force, N; D_p is the punch diameter, mm; D_b is the blank diameter, mm; t is the sheet metal (blank) thickness, mm; and S_{ut} is the tensile strength of the sheet metal, MPa.

TABLE 7.3

Conditions for DD Feasibility

DD Parameter	DR	Reduction, r	t/D_b
Condition	<2	<0.5	>1%

Holding Force: The holding force in drawing is usually one-third of the drawing force; hence, it can be found as follows:

$$F_h = 0.33\, F_{dd} \tag{7.16}$$

where F_h is the holding force in DD; and F_{dd} is the DD force.

7.6 Roll Bending and Roll Forming

Roll bending involves the forming of large sheet metal parts into curved sections by passing them between rolls. Roll bending has an advantage over bending because no cutting of sheet metal is required and no material is wasted. It is used if a metal sheet panel is to be given a larger smooth radius.

Roll forming involves continuous bending of a large sheet metal by using opposing rolls to produce long sections of formed shapes. Roll forming is an ideal process for making components with a consistent profile running the entire length of the part, in which coiled sheet metal is passed through a horizontal series of rotating dies to produce a uniform cross-sectional shape. In roll forming, each set of rolling dies makes an incremental change to the sheet metal until the desired profile is achieved. Finally, the bended piece is cut to the specified length of the final part.

7.7 Problems and Solutions—Examples in Sheet Metal Forming

7.7.1 Examples in Shearing, Blanking, and Piercing

EXAMPLE 7.1: COMPUTING CLEARANCE, PUNCH DIAMETER AND DIE DIAMETER

A round disk of 140 mm diameter is to be *blanked* from a 3.3 mm thick soft stainless steel strip. Determine the following: (a) clearance, (b) punch diameter, and (c) die diameters.

Solution

$D_b = 140$ mm, $t = 3.3$ mm.

By reference to Table 7.1, $A_c = 0.06$.

(a) By using Equation 7.1,

$$c = A_c\,t = 0.06 \times 3.3 = 0.198 \text{ mm} \sim 0.2 \text{ mm}$$

Clearance $= 0.2$ mm.

(b) By using Equation 7.3,

$$\text{Blanking punch diameter} = D_b - 2c = 140 - (2 \times 0.2)$$
$$= 140 - 0.4 = 139.6 \text{ mm}$$

(c) By using Equation 7.4,

$$\text{Blanking die diameter} = D_b = 140 \text{ mm}$$

EXAMPLE 7.2: CALCULATING SHEARING FORCE

By using the data in Example 7.1, calculate the shearing force for blanking if the tensile strength of the soft stainless steel is 560 MPa.

Solution

$t = 3.3$ mm, $S_{ut} = 560$ MPa, $D_b = 140$ mm.

$L = \text{Length of cut} = \text{Circumference} = \pi D_b = 3.142 \times 140 = 439.8$ mm.

By using Equation 7.2,

$$F = \tau \cdot t \cdot L = 0.7\,S_{ut}\,t \cdot L = 0.7 \times 560 \times 3.3 \times 439.8 = 568{,}955 \text{ N} = 569 \text{ kN}$$

Shearing or blanking force $= 569$ kN.

EXAMPLE 7.3: CALCULATING PIERCING PUNCH DIAMETER AND PIERCING DIE DIAMETER

A 10 mm hole is punched (pierced) in a 3 mm thick cold rolled (CR) steel strip.

Calculate the following: (a) clearance, (b) punch diameter, and (c) die diameter.

Solution

$D_h = 10$ mm, $t = 3$ mm.

By reference to Table 7.1, $A_c = 0.075$.

(a) By using Equation 7.1,

$$c = A_c\,t = 0.075 \times 3 = 0.225 \text{ mm}$$

(b) By using Equation 7.5,

$$\text{Piercing punch diameter} = D_h = 10 \text{ mm}$$

(c) By using Equation 7.6,

$$\text{Piercing die diameter} = D_h + 2c = 10 + (2 \times 0.225)$$
$$= 10 + 0.45 = 10.45 \text{ mm}$$

Die diameter = 10.45 mm.

EXAMPLE 7.4: CALCULATING THE PUNCH AND DIE SIZES FOR PRODUCING A WASHER

A large washer is to be produced by blanking and piercing out of a soft stainless steel sheet stock of 3.5 mm thickness by using a compound die. The outside diameter of the washer is 55 mm, and the inside diameter is 20 mm. Determine (a) the punch and die sizes for the blanking operation and (b) the punch and die sizes for the piercing.

Solution

In order to produce the washer, the first step is blanking, and then piercing.

$D_b = 55$ mm, $D_h = 20$ mm, $t = 3.5$ mm.

By reference to Table 7.1 (for soft stainless steel), $A_c = 0.06$.

By using Equation 7.1,

$$c = A_c\, t = 0.06 \times 3.5 = 0.21 \text{ mm}$$

(a) By using Equations 7.3 and 7.4,

$$\text{Blanking punch diameter} = D_b - 2c = 55 - (2 \times 0.21) = 54.58 \text{ mm}$$

$$\text{Blanking die diameter} = D_b = 55 \text{ mm}$$

Blanking punch diameter = 54.58 mm; blanking die diameter = 55 mm.

(b) By using Equations 7.5 and 7.6,

$$\text{Piercing punch diameter} = D_h = 20 \text{ mm}$$

$$\text{Piercing die diameter} = D_h + 2c = 20 + (2 \times 0.21) = 20 + 0.42 = 20.42 \text{ mm}$$

Piercing punch diameter = 20 mm; piercing die diameter = 20.42 mm.

EXAMPLE 7.5: CALCULATING THE PRESS FORCE REQUIRED FOR A COMPOUND DIE

By using the data in Example 7.4, calculate the minimum press force required to produce the washer by using the compound die.

Solution

By reference to Example 7.2,

Shear strength of soft stainless steel $= \tau = 0.7\,S_{ut} = 0.7 \times 560 = 392$ MPa

For *blanking*: Length of cut = blank circumference $= L_b = \pi D_b = 55\pi = 172.78$ mm.

For *piercing*: Length of cut = blank circumference $= L_h = \pi D_h = 20\pi = 62.8$ mm.

By using Equation 7.2,

Blanking force $= \tau \cdot t \cdot L_b = 392 \times 3.5 \times 172.78 = 237{,}064.58$ N $= 237.1$ kN

Piercing force $= \tau \cdot t \cdot L_h = 392 \times 3.5 \times 62.8 = 86{,}205.3$ N $= 86.2$ kN

$$\text{Press (total) Force} = \text{Blanking force} + \text{Piercing force}$$
$$= 237.1 + 86.2 = 323.3 \text{ kN}$$

The minimum press force required to produce the washer $= 323.3$ kN.

7.7.2 Examples in Bending

EXAMPLE 7.6: CALCULATING THE BLANK SIZE FOR BENDING

A 6061 aluminum alloy sheet blank is to be bent as shown in Figure E7.6. Determine the starting blank size.

Solution

$\alpha' = 120°$, $t = 3$ mm, $2t = 6$ mm, $R = 4.8$ mm, $w = 45$ mm.

Bend angle $= \alpha = 180 - \alpha' = 180 - 120 = 60°$.

Since $R < 2t$, $K_{ba} = 0.33$ (see Section 7.4.2).

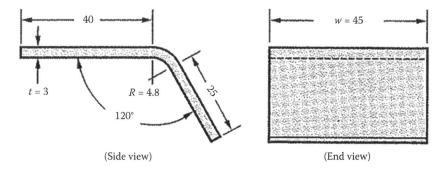

(Side view)   (End view)

FIGURE E7.6
The bent sheet metal (all dimensions in mm).

By using Equation 7.9,

$$A_b = [2\pi\ \alpha\ (R + K_{ba}\ t)]/360 = \{120\ \pi\ [4.8 + (0.33)\ (3)]\}/360 = 6.06\ \text{mm}$$

By reference to Figure E7.6, $L = 40 + 25 = 65$ mm.

By using Equation 7.10,

$$L_{blank} = L + A_b = 65 + 6.06 = 71.1\ \text{mm}$$

Starting blank size (length) $= 71.1$ mm.

EXAMPLE 7.7: CALCULATING THE BENDING FORCE

The 6061 aluminum alloy has an ultimate tensile strength of 310 MPa. By using the data in Example 7.6, compute the bending force if a V-die is used with a die opening dimension of 25 mm.

Solution

$t = 3$ mm, $w = 45$ mm, $D = 25$ mm, $S_{ut} = 310$ MPa.

By reference to Table 7.2, $K_{bf} = 1.33$ (for V-bending).

By using Equation 7.8,

$$F = (K_{bf}\ S_{ut}\ w\ t^2)/D = (1.33 \times 310 \times 45 \times 3^2)/25 = 6679.3\ \text{N} = 6.67\ \text{kN}$$

The bending force $= 6.67$ kN.

7.7.3 Examples in Deep Drawing

EXAMPLE 7.8: CALCULATING THE CLEARANCE FOR *DD*

A sheet metal stock of thickness 1.8 mm is to be deep drawn to produce a cup. What clearance should be allowed?

Solution

$t = 1.8$ mm, $c = ?$

By using Equation 7.11,

$$c = 1.1\ t = 1.1 \times 1.8 = 1.98\ \text{mm}$$

Clearance $= 2$ mm.

EXAMPLE 7.9: CALCULATING DRAWING RATIO, REDUCTION, AND t/D_b RATIO FOR *DD*

A *DD* operation is used to form a cylindrical cup with inside diameter $(ID) = 78$ mm and height $= 55$ mm. The starting blank diameter $= 140$ mm and the stock thickness $= 2.7$ mm. Calculate the following: (a) DR, (b) reduction, and (c) t/D_b.

Solution

$D_b = 140$ mm, $t = 2.7$ mm, $DR = ?$, $r = ?$, $t/D_b = ?$

By reference to Figure 7.4, $D_p = $ cup $ID = 78$ mm.

(a) By using Equation 7.12,

$$DR = D_b/D_p = (140)/(78) = 1.79$$

(b) By using Equation 7.13,

$$r = (D_b - D_p)/D_b = (140 - 78)/140 = 0.44$$

(c) By using Equation 7.14,

$$t/D_b = (2.7)/(140) = 0.019 = 1.9\%$$

$DR = 1.8$; reduction $= 0.44$; *thickness-to-diameter ratio* $= 1.9\%$.

EXAMPLE 7.10: DETERMINING THE FEASIBILITY OF *DD* OPERATION

By using the data in Example 7.9, determine whether or not the *DD* operation is feasible.

Solution

$DR = 1.8 < 2$; $r = 0.44 < 0.5$; $t/D_b = 1.9\% > 1\%$.

By reference to Table 7.3, all conditions for *DD* feasibility are satisfied. Hence, the *DD* operation is feasible.

EXAMPLE 7.11: COMPUTING THE *DD* FORCE AND HOLDING FORCE

By using the data in Example 7.9, compute (a) the force required to deep draw and (b) holding force. The sheet metal is soft stainless steel.

Solution

$D_b = 140$ mm, $t = 2.7$, $D_p = 78$ mm.

By reference to Example 7.2, the tensile strength of soft stainless steel $= S_{ut} = 560$ MPa.

(a) By using Equation 7.15,

$$F_{dd} = \pi D_p\, t\, S_{ut}\, [(D_b/D_p) - 0.7] = (\pi \times 78 \times 2.7 \times 560)\, [(140/78) - 0.7]$$
$$= 405{,}657.44\ \text{N}$$

Deep drawing force $= 405.66$ kN.

(b) By using Equation 7.16,

$$F_h = 0.33\ F_{dd} = 0.33 \times 405.66 = 133.8\ \text{kN}$$

Holding force $= 133.8$ kN.

Questions and Problems

7.1 (a) List at least four application sectors of metal plates and sheets. (b) Illustrate an open shearing operation with the aid of a sketch.

7.2 Differentiate between the following terms: (a) metal plate and metal sheet, (b) compound die and progressive die.

7.3 Differentiate between the following terms with the aid of diagrams: (a) blanking and piercing, (b) V-bending and edge bending.

7.4 Draw a labeled diagram showing the following geometric design features in blanking/piercing: clearance, punch radius, sheet thickness, and punch force.

7.5 (a) Define the term: *bending*. (b) Draw a labeled diagram showing the following geometric design features in bending: bend radius, included angle, bend angle, sheet width, and sheet thickness

7.6 (a) Define the term *deep drawing*. (b) Draw a diagram showing the following design features in *DD*: punch diameter, punch corner radius, die corner radius, blank diameter, and clearance.

P7.7 A round disk of 130 mm diameter is to be blanked from a 2.8 mm thick 1100S aluminum alloy strip. Determine (a) clearance, (b) punch diameter, and (c) die diameter.

P7.8 A square disk of 100 mm side length is to be blanked from a 3 mm thick metal strip having a tensile strength of 560 MPa. Compute the blanking force.

P7.9 A 9 mm hole is punched (pierced) in a 2.7 mm thick soft stainless steel strip. Calculate the following: (a) clearance, (b) punch diameter, and (c) die diameter.

P7.10 A large washer is to be produced by blanking and piercing out of a *CR* steel sheet stock of 2.8 mm thickness by using a compound die. The washer's outside diameter is 57 mm, and the inside diameter is 22 mm. Compute the punch and die sizes for (a) the blanking operation and (b) piercing.

P7.11 By using the data in Problem 7.10, calculate the minimum press force required to produce the washer by using the compound die, if the tensile strength of *CR* steel is 420 MPa.

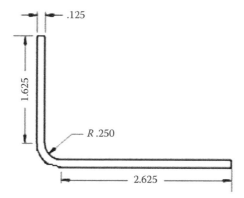

FIGURE P7.12
The bent sheet metal (all dimensions in inches).

P7.12 An aluminum alloy sheet blank is to be bent as shown in Figure P7.12. Calculate the starting blank size.

P7.13 By using the data in Figure P7.12, compute the bending force if a die opening dimension of 28 mm is used. The aluminum alloy has an ultimate tensile strength of 350 MPa; the sheet width is 43 mm.

P7.14 A *DD* operation is used to form a cylindrical cup with inside diameter of 73 mm and height = 52 mm. The starting blank diameter = 133 mm, and the stock thickness = 2.3 mm. Calculate the following: (a) drawing ratio, (b) reduction, and (c) t/D_b.

References

Altan, T., Tekkaya, A.E. 2012. *Sheet Metal Forming: Fundamentals.* Materials Park, OH: ASM International.

Gisario, A., Barletta, M., Venettassi, S. 2016. Improvements in springback control by external force laser-assisted sheet bending of titanium and aluminum alloys. *Optics and Laser Technology*, 86(12), 46–53.

Groover, M.P. 2014. *Fundamentals of Modern Manufacturing.* New York: John Wiley and Sons.

Huda, Z. 2017. *Materials Processing for Engineering Manufacture.* Switzerland: Trans Tech.

8

Machining I—Mechanics of Machining

8.1 Mechanics of Chip Formation—Fundamentals

Chip formation is a localized shear process in a narrow region where the material is plastically deformed and slides off along the rake face of the cutting edge. Chip formation affects the surface finish, cutting forces, temperature, tool life, and dimensional tolerance (Liang and Shih, 2016). A chip consists of two sides: (a) the *shiny side* (flat and uniform side) that is in contact with the tool, due to frictional effects; and (b) the *jagged side* that is the free workpiece surface, due to shear. The main cutting action in machining involves shear deformation of the work material followed by material removal in the form of a chip.

8.2 Orthogonal Cutting and Oblique Cutting

The distinction between orthogonal cutting and oblique cutting is illustrated in Table 8.1.

8.3 Mathematical Modeling of Mechanics of Chip Formation

Figure 8.1 illustrates the various chip formation parameters in orthogonal machining.

The chip thickness ratio, r, is given by (Huda, 2017)

$$r = t_0/t_c \tag{8.1}$$

where t_0 is chip thickness before cut, and t_c is chip thickness after cut.

The shear plane angle, ϕ, can be found by the formula

$$\tan\phi = (r\cos\alpha)/(1 - r\sin\alpha) \tag{8.2}$$

where α is the rake angle, and ϕ is the shear plane angle.

TABLE 8.1

Orthogonal and Oblique Cutting

Number	Orthogonal Cutting	Oblique Cutting
1	The cutting edge is perpendicular to the cutting velocity vector.	The cutting edge is inclined at an angle to the cutting velocity vector.
2	It is a two-dimensional machining.	It is a three-dimensional machining.
3	Examples: Straight turning, plain milling, surface broaching, etc.	Example: Oblique turning

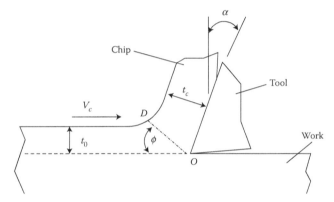

FIGURE 8.1
Chip formation parameters in orthogonal machining (V_c = cutting speed).

The shear strain, γ, can be determined as follows:

$$\gamma = \tan(\phi - \alpha) + \cot \phi \qquad (8.3)$$

It is evident from Figure 8.1 that

$$OD = t_0/\sin \phi = (t_c)/[\cos(\phi - \alpha)] \qquad (8.4)$$

By combining Equation 8.1 and Equation 8.4, we obtain

$$r = \sin \phi/\cos(\phi - \alpha) \qquad (8.5)$$

8.4 Types of Chips in Machining

There are four types of chips in machining: (a) discontinuous or segmented chip, (b) continuous chip, (c) continuous chip with built-up edge (BUE), and

(d) serrated chip. The formation of a particular type of chip depends on the type of material being machined and the cutting conditions of the machining operation. Discontinuous chips break up into small segments and are convenient to collect, handle, and dispose of. Discontinuous chips tend to be formed when machining a brittle material (e.g., cast iron, bronze, etc.), using a large chip thickness, or machining at a low cutting speed.

Continuous chips are formed by the continuous plastic deformation of metal without fracture in front of the cutting edge of the tool when machining at high cutting speed or when the work material is ductile. Mild steel and copper are considered to be the most desirable materials for forming continuous chips.

8.5 Forces in Machining and Their Analysis/Mathematical Modeling

The knowledge of *forces in machining* is important due to the following reasons: (a) estimation of power requirements and the selection of machine tool electric motor; (b) machine tool design (e.g., static/dynamic stiffness); (c) achievement of the required dimensional accuracy; and (d) selection of appropriate tool holders (Walker, 2004).

There are two forces acting on a tool during machining: (1) cutting force, F_c, and (2) thrust force, F_t. The *cutting force* (F_c) acts in the direction of the cutting speed (V_c), whereas the thrust force (F_t) acts in a direction normal to the cutting force (see Figure 8.2). The forces F_c and F_t can be measured by use of the device: dynamometer.

Figure 8.2 illustrates four forces acting on the chip/work during machining: (1) friction force, F; (2) normal force, N; (3) shear force, F_s; and (4) normal force to shear plane, N_s.

The friction force is related to the normal force by

$$F = \mu N \qquad (8.6)$$

where μ is the coefficient of friction between the tool–chip interface and can be measured by

$$\mu = \tan \beta \qquad (8.7)$$

The shear stress can be determined by

$$\tau = F_s/A_s \qquad (8.8)$$

where τ is the shear stress (=shear strength of the work material), MPa; F_s is the shear force, N; and A_s is the shear plane area, mm^2.

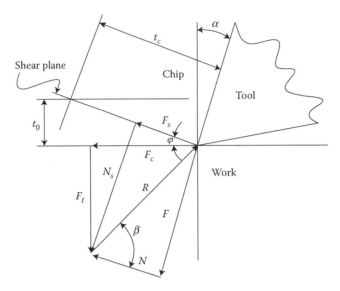

FIGURE 8.2
Forces acting on tool and chip/work during machining (F_c = cutting force, F_t = thrust force, F = friction force, N = normal force, F_s = shear force, N_s = normal force to shear plane).

The shear plane area, A_s, can be determined by

$$A_s = (t_0 \, w)/(\sin \phi) \tag{8.9}$$

where w is the work width; t_0 is chip thickness before cut; and ϕ is the shear plane angle.

It has been stated in the preceding section that the cutting force (F_c) and the thrust force, (F_t) can be measured by dynamometer. Once these two forces are known, the four forces (acting on chip) can be determined by the following mathematical relationships (see Figure 8.2):

$$F = F_c \sin \alpha + F_t \cos \alpha \tag{8.10}$$

$$N = F_c \cos \alpha - F_t \sin \alpha \tag{8.11}$$

$$F_s = F_c \cos \phi - F_t \sin \phi \tag{8.12}$$

$$N_s = F_c \sin \phi + F_t \cos \phi \tag{8.13}$$

8.5.1 Power in Machining

The total input power to cutting can be given by

$$P_c = F_c \, v \tag{8.14}$$

where P_c is the cutting power, W; F_c is the cutting force, N; and v is the cutting speed, m/s.

Another useful term is the *unit power*, which is defined as "power per unit volume rate of metal cut"; that is, unit power can be numerically defined as

$$P_u = P_c/MRR \qquad (8.15)$$

where P_u is the unit power, $W \cdot s/mm^3$; and MRR is the material removal rate, mm^3/s. The unit power is a material property—a handbook value.

8.5.2 Effect of Shear Plane Angle on Machining

It can be geometrically shown that an increase in the shear plane angle (ϕ) results in lowering the shear plane area (A_s) (see Figures 8.1 and 8.2). Since the shear stress (τ) is applied across this area, the shear force ($F_s = A_s\tau$) required to form the chip will decrease, which means easier machining. Hence, a greater shear plane angle results in lower cutting force, lower power requirements, and lower cutting temperature.

8.6 Merchant's Equation

Merchant's equation enables us to express the shear plane angle (ϕ) in terms of rake angle (α) and friction angle (β), as follows:

$$\phi = 45° + (\alpha/2) - (\beta/2) \qquad (8.16)$$

Equation 8.14 is of great technological importance in the machining of metals and was derived by Eugene Merchant (Merchant, 1945). It was established in the preceding section that a higher shear plane angle (ϕ) results in easier machining. According to Merchant's equation, in order to increase the shear plane angle, one must either increase the rake angle, α, or reduce the friction angle, β (or coefficient of friction by using a lubricant cutting fluid) (see Equation 8.7).

8.7 Problems and Solutions—Examples in Mechanics of Machining

EXAMPLE 8.1: COMPUTING CHIP THICKNESS RATIO, SHEAR PLANE ANGLE, AND SHEAR STRAIN

In an orthogonal machining operation, the cutting tool has a rake angle of 12°. The chip thicknesses before and after the cut are 0.55 mm and 1.2 mm, respectively. Calculate the following: (a) chip thickness ratio, (b) shear plane angle, and (c) shear strain.

Solution

$\alpha = 12°$, $t_0 = 0.55$ mm, $t_c = 1.2$ mm, $r = ?$, $\phi = ?$, $\gamma = ?$

(a) By using Equation 8.1,

$$r = t_0/t_c = (0.55)/(1.2) = 0.46$$

Chip thickness ratio $= 0.46$.

(b) By using Equation 8.2,

$$\tan \phi = (r \cos \alpha)/(1 - r \sin \alpha) = (0.46 \times \cos 12°)/[1 - (0.46)(\sin 12°)]$$

$$\tan \phi = (0.45)/(1 - 0.095) = (0.45)/(0.9) = 0.5$$

$$\phi = 26.5°$$

Shear plane angle $= 26.5°$.

(c) By using Equation 8.3,

$$\gamma = \tan(\phi - \alpha) + \cot \phi = \tan(26.5° - 12°) + \cot 26.5° = 0.26 + 2 = 2.26$$

Shear strain $= 2.26$.

EXAMPLE 8.2: VERIFYING THE FORMULA FOR CHIP THICKNESS RATIO—EQUATION 8.5

By using the data in Example 8.1, verify the validity of Equation 8.5.

Solution

$\alpha = 12°$, $r = 0.46$, $\phi = 26.5°$.

By using Equation 8.5

$$r = \sin \phi/\cos(\phi - \alpha) = (\sin 26.5°)/[\cos(26.5° - 12°)]$$
$$= (0.446)/(0.97) = 0.46$$

Hence, Equation 8.5 is verified by using the data in Example 8.1.

EXAMPLE 8.3: CALCULATING SHEAR STRENGTH OF A MATERIAL BASED ON MACHINING DATA

In an orthogonal machining operation, the cutting force and thrust force are measured as 1650 N and 1340 N, respectively. Compute the shear strength of the work material, based on the following machining data: *rake angle* $= 11°$, *shear plane angle* $= 26°$, chip thickness before cut $= 0.6$ mm, and width of the cutting operation $= 3.2$ mm.

Solution

$F_c = 1650$ N, $F_t = 1340$ N, $\alpha = 11°$, $\phi = 26°$, $t_0 = 0.6$ mm, $w = 3.2$ mm, $\tau = ?$

By using Equation 8.12,

$$F_s = F_c \cos \phi - F_t \sin \phi = (1650 \times \cos 26°) - (1340 \times \sin 26°) = 895.5 \text{ N}$$

By using Equation 8.9,

$$A_s = (t_0 \, w)/(\sin \phi) = (0.6 \times 3.2)/(\sin 26°) = (1.92)/(0.438) = 4.38 \text{ mm}^2$$

By using Equation 8.8,

$$\tau = F_s/A_s = 895.5/4.38 = 204.4 \text{ MPa}$$

The shear strength of the work material $= 204.4$ MPa.

EXAMPLE 8.4: COMPUTING THE FRICTION ANGLE AND THE COEFFICIENT OF FRICTION

Based on the data in Example 8.3, calculate the following: (a) the friction angle and (b) the coefficient of friction for the machining operation.

Solution

$\alpha = 11°$, $\phi = 26°$, $\beta = ?$, $\mu = ?$

(a) By using Equation 8.16,

$$\phi = 45° + (\alpha/2) - (\beta/2)$$
$$26 = 45 + (11/2) - (\beta/2)$$
$$(\beta/2) = 45 + (11/2) - 26 = 24.5$$
$$\beta = 49°$$

The friction angle $= 49°$.

(b) By using Equation 8.7,

$$\mu = \tan \beta = \tan 49° = 1.15$$

The coefficient of friction $= 1.15$.

EXAMPLE 8.5: THE APPLICATION OF MERCHANT'S FORMULA

In an orthogonal cutting, the chip thickness ratio is 0.35, the rake angle is 6°, the depth of the cut (chip thickness before cut) is 0.3 mm, and the width of the cutting operation is 5 mm. Compute (a) the chip thickness after the cut, (b) the shear plane angle, (c) the friction angle, (d) the coefficient of friction, and (e) the shear strain.

Solution

$r = 0.35$, $\alpha = 6°$, $t_0 = 0.3$ mm, $w = 5$ mm, $t_c = ?$, $\phi = ?$, $\beta = ?$, $\mu = ?$, $\gamma = ?$

(a) By using Equation 8.1,

$$t_c = t_0/r = 0.3/0.35 = 0.86 \text{ mm}$$

Chip thickness after cut $= 0.86$ mm.

(b) By using Equation 8.2,

$$\tan \phi = (r \cos \alpha)/(1 - r \sin \alpha) = (0.35 \times \cos 6°)/[1 - (0.35)(\sin 6°)]$$

$$\tan \phi = (0.35)/(1 - 0.036) = 0.35/0.963 = 0.363$$

$$\phi = \tan^{-1} 0.363 = 20°$$

Shear plane angle $= 20°$.

(c) By using Equation 8.16,

$$\phi = 45° + (\alpha/2) - (\beta/2)$$

$$20 = 45 + (6/2) - (\beta/2)$$

$$\beta = 56°$$

The friction angle $= 56°$.

(d) By using Equation 8.7,

$$\mu = \tan \beta = \tan 56° = 1.48$$

The coefficient of friction $= 1.48$.

(e) By using Equation 8.3,

$$\gamma = \tan(\phi - \alpha) + \cot \phi = \tan(20° - 6°) + \cot 20° = 0.25 + 2.74 = 3$$

Shear strain $= 3$.

EXAMPLE 8.6: COMPUTING THE FRICTION FORCE AND NORMAL FORCE

In an orthogonal cutting operation, the rake angle is $-5°$, the cutting force is 1450 N, and the thrust force is 1000 N. Compute the (a) friction force and (b) normal force.

Solution

$\alpha = -5°$, $F_c = 1450$ N, $F_t = 1000$ N, $F = ?$, $N = ?$

(a) By using Equation 8.10,

$$F = F_c \sin \alpha + F_t \cos \alpha = (1450 \times \sin -5°) + (1000 \times \cos -5°) = 869.9 \text{ N}$$

The friction force $= 870$ N.

(b) By using Equation 8.11,

$$N = F_c \cos \alpha - F_t \sin \alpha = (1450 \times \cos -5°) - (1000 \times \sin -5°)$$

$$N = 1444.5 - (-87.1) = 1531.6 \text{ N}$$

The normal force = 1531.6 N.

EXAMPLE 8.7: CALCULATING THE NORMAL FORCE TO SHEAR PLANE

By using the data in Example 8.3, calculate the normal force to shear plane.

Solution

$F_c = 1650$ N, $F_t = 1340$ N, $\phi = 26°$, $N_s = ?$

By using Equation 8.13,

$$N_s = F_c \sin \phi + F_t \cos \phi = (1650 \times \sin 26°) + (1340 \times \cos 26°)$$

$$N_s = 723.3 + 1204.4 = 1927.7 \text{ N}$$

The normal force to shear plane = 1927.7 N.

EXAMPLE 8.8: COMPUTING THE CHIP THICKNESS RATIO WHEN THE COEFFICIENT OF FRICTION IS GIVEN

In orthogonal cutting, the rake angle is 28° and the coefficient of friction is 0.3. Determine the (a) shear plane angle and (b) the chip thickness ratio.

Solution

$\alpha = 28°$, $\mu = 0.3$, $\phi = ?$, $r = ?$

By using Equation 8.7,

$$\mu = \tan \beta$$

$$0.3 = \tan \beta$$

$$\beta = \tan^{-1} 0.3 = 16.7° \sim 17°$$

(a) By using Equation 8.16,

$$\phi = 45° + (\alpha/2) - (\beta/2) = 45 + (28/2) - (17/2) = 50.5°$$

The shear plane angle $= \phi = 50.5°$.
(b) By using Equation 8.5,

$$r = \sin \phi / \cos (\phi - \alpha) = (\sin 50.5°)/[\cos (50.5° - 28°)]$$
$$= (0.77)/(0.92) = 0.84$$

The chip thickness ratio $= r = 0.84$.

EXAMPLE 8.9: DETERMINING THE EFFECT OF DOUBLING FRICTION COEFFICIENT ON SHEAR PLANE ANGLE

By using the data in Example 8.8, compute the shear plane angle if the coefficient of friction is doubled.

Solution

$\alpha = 28°, \mu = 0.3$

For Case 1, $\mu = 0.3 = \tan \beta$, or $\beta = 17°$, and hence, by using Equation 8.16, we obtain

$$\phi = 45° + (\alpha/2) - (\beta/2) = 50.5°$$

For Case 2, the friction coefficient is doubled: $\mu = 2 \times 0.3 = 0.6$.
By using Equation 8.7,

$$\mu = \tan \beta$$

$$0.6 = \tan \beta$$

$$\beta = \tan^{-1} 0.6 = 31°$$

By using Equation 8.16,

$$\phi = 45° + (\alpha/2) - (\beta/2) = 45 + (28/2) - (31/2) = 45 + 14 - 15.5 = 43.5°$$

Hence, the shear plane angle is reduced from 50.5° to 43.5° if the friction coefficient is doubled.

EXAMPLE 8.10: DETERMINING THE POWER INPUT FOR MACHINING

In a turning operation for machining stainless steel, the cutting speed is 16 m/min and the cutting force is measured to be 500 N. What is the value of the cutting power?

Solution

$F_c = 500$ N, $v = 16$ m/min $= 16/60$ m/s $= 0.27$ m/s.
By using Equation 8.14,

$$P_c = F_c \, v = 500 \times 0.27 = 133.3 \text{ W}$$

The cutting power $= P_c = 133.3$ W.

EXAMPLE 8.11: DETERMINING THE MATERIAL REMOVAL RATE WHEN *d* AND *f* ARE UNKNOWN

By using the data in Example 8.10, calculate the material removal rate in the machining.

Solution

$P_c = 133.3$ W; For stainless steel, $P_u = 4$ W · s/mm^3.

P_u is the unit power, W · s/mm^3; and *MRR* is the material removal rate, mm^3/s.

By using Equation 8.15,

$$P_u = P_c/MRR$$

or $MRR = P_c/P_u = 133.3/4 = 33.33$ mm^3/s $= 2000$ mm^3/min

The material removal rate $= 2000$ mm^3/min.

Questions and Problems

8.1 List the three fundamental machining parameters and define them.

8.2 Briefly explain the mechanics of chip formation with the aid of a diagram.

8.3 Draw diagrams showing the four types of chips in machining.

8.4 Apply Merchant's equation to explain that the use of lubricants results in easier machining.

P8.5 Calculate the shear strength of the work material based on the following orthogonal machining operation's data: *cutting force* $= 1700$ N, *thrust force* $= 1440$ N, *rake angle* $= 13°$, *shear plane angle* $= 23°$, chip thickness before cut $= 0.7$ mm, and width of the cutting operation $= 3.7$ mm.

P8.6 Based on the data in P8.5, calculate the following: (a) the friction angle and (b) the coefficient of friction for the machining operation.

P8.7 In an orthogonal machining operation, the chip thickness before and after the cut are 0.48 mm and 0.7 mm, respectively. The cutting tool has a rake angle of 13°. Calculate the following: (a) chip thickness ratio, (b) shear plane angle, and (c) shear strain.

P8.8 In an orthogonal cutting operation, the rake angle is –7°, the cutting force is 1600 N, and the thrust force is 1080 N. Compute the (a) friction force and (b) normal force.

P8.9 In orthogonal cutting, the rake angle is 25° and the coefficient of friction is 0.4. Determine the (a) shear plane angle and (b) chip thickness ratio.

P8.10 By using the data in P8.9, compute the shear plane angle if the coefficient of friction is tripled.

References

Huda, Z. 2017. *Materials Processing for Engineering Manufacture*. Switzerland: Trans Tech.

Liang, S.Y., Shih, A.J. 2016. *Analysis of Machining and Machine Tools*, 1st Edition. Berlin, Germany: Springer International.

Merchant, M.E. 1945. Mechanics of the metal cutting process: II. Plasticity conditions in orthogonal cutting. *Journal of Applied Physics*, 16, 318–324.

Walker, J.R. 2004. *Machining Fundamentals*, 8th Edition. Tinley Park, IL: Goodheart-Willcox.

9

Machining II—Cutting Tools, Turning, and Drilling

9.1 Cutting Tools and Tool Life

9.1.1 Single-Point and Multiple-Point Cutting Tools

Based on cutting edge, cutting tools may be divided into two groups: (a) single-point cutting tools (having one cutting edge) and (b) multiple-point cutting tools (having multiple edges).

Examples of single-point cutting tools include the tools used in turning, shaping, planing, etc. (see Figure 9.1[a]). Examples of multiple-point cutting tools include tools used in milling, drilling, grinding, etc. (see Figure 9.1[b]).

9.1.2 Cutting Tool Materials

The success of a machining operation strongly depends on the performance of the cutting tool material. A cutting tool material must be harder than the workpiece material. The basic requirements of a cutting tool material include (1) hardness, (2) hot hardness, (3) toughness, and (4) wear resistance (Huda, 2017). Hot hardness is important so as to enable the cutting tool to withstand the heat generated during machining, Toughness of cutting tools is needed so that tools do not chip or fracture, especially during interrupted cutting operations.

A variety of cutting tool materials are used in machining. High carbon tool steel is the oldest tool material. There has been significant development in cutting tool materials, which are arranged in increasing order of high performance as follows: high-speed steel (HSS), high molybdenum HSS, high tungsten HSS, cemented carbides (WC-Co composites), coated carbides, ceramics (fine-grained Al_2O_3, pressed and sintered), cubic boron nitride (CBN), and synthetic diamonds. The choice of a cutting tool depends on the cutting speed, type of machining operations, and work material. For example, machining of hard materials at high cutting speeds requires the use of ceramic tool, CBN, or diamond. For machining of softer materials at low or moderate cutting speeds, the use of HSS or cemented carbide tool is recommended.

(a)   (b)

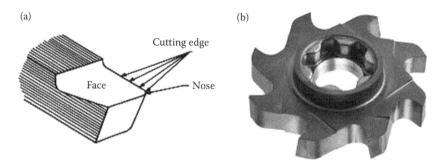

FIGURE 9.1
Cutting tools: (a) single-point cutting tool, (b) multiple-point cutting tool.

9.1.3 Tool Life and Its Mathematical Modeling—Taylor's Equation

All cutting tools wear with machining. When the wear exceeds a limit, the tool is said to have failed. Hence, tool wear is an important limitation to machining productivity and part quality (Drouillet et al., 2016). *Tool life* is the total service time (in minutes or seconds) of a cutting tool to failure at certain cutting speed and other service conditions. The life of a cutting tool can significantly be enhanced by use cutting fluid during the machining operation (Youssef and El-Hofy, 2008). There are two main types of cutting fluids: (1) *coolants*—designed to reduce effects of heat in machining, and (2) *lubricants*—designed to reduce tool chip and tool work friction.

Tool life strongly depends on the cutting speed, and the mathematical relationship is expressed by Taylor's equation, as follows:

$$v \, T_L^n = C \tag{9.1}$$

where v is the cutting speed, m/min; and T_L is the tool life, min. By taking logarithm of both sides of Equation 9.1, we obtain

$$\log v + n \log T_L = \log C \tag{9.2}$$

or

$$\log v = -n \log T_L + \log C \tag{9.3}$$

Equation 9.3 clearly indicates its slope-intercept form; here n is the slope of the ($\log T_L - \log v$) graphical plot; and C is the intercept on the $\log v$ axis at one minute tool life (see Figure 9.2).

The knowledge of tool life, T_L, is of great technological importance to a manufacturing engineer. A high exponent n value (in the Taylor's equation) indicates a high tool life and better performance of the cutting tool material. The ranges of n and C values (in the Taylor's equation) for various cutting tool materials are presented in Table 9.1.

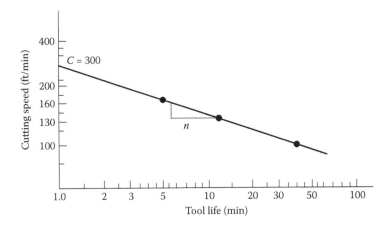

FIGURE 9.2
The log T_L – log v graphical plot representing Equation 9.3.

TABLE 9.1

The n and C Value Ranges in Taylor's Equation

Cutting-Tool Material	HSS	Cast Alloys	Cemented Carbides	Coated Carbides	Ceramics
N	0.08–0.2	0.1–0.15	0.2–0.5	0.4–0.6	0.5–0.7
C, m/min (for steel cutting)	70	n/a	500	700	3000

9.2 Machining Operation—Turning

9.2.1 Machining Parameters and Turning Operation

In machining, a relative motion between the tool and the work is required. The primary motion is called *cutting speed*, whereas the secondary motion is called *feed*. In a machining operation, there are three fundamental machining parameters: cutting (surface) speed (v), feed (f), and depth of cut (d). *Turning* is one of the most basic machining operations. *Turning* involves the removal of metal from the outer diameter of a rotating cylindrical workpiece (Figure 9.3).

There are various types of turning operations, including straight turning, facing, contour turning, form turning, chamfering, cut-off turning, boring, and threading (Huda, 2017). In straight turning, the cutting tool moves longitudinally to produce a straight workpiece (Figure 9.3).

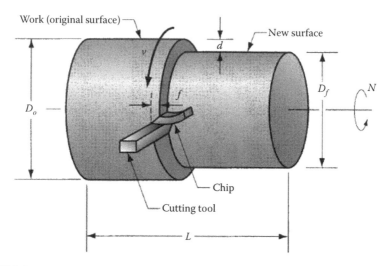

FIGURE 9.3
Turning operation's geometric parameters.

9.2.2 Mathematical Modeling for Turning

In *turning*, the work diameter is reduced from the original diameter, D_o, to the final diameter, D_f, along the length of the cut, L (see Figure 9.3).

The cutting speed (surface speed) of the workpiece can be found as follows:

$$v_{max} = \pi \, D_o \, N \tag{9.4}$$

where v_{max} is the maximum cutting speed, mm/min; D_o is the original diameter of the work, mm; and N is the spindle rpm or the rotational speed of the work, rev/min.

The maximum cutting speed and feed strongly depend on the work material and the tool (see Table 9.2). It is already mentioned in Section 9.1.2 that a ceramic tool should be used for machining hard materials at high cutting speed, whereas an HSS tool is used for softer materials.

The average cutting speed of the work is given by

$$v_{av} = \pi \, D_{av} \, N \tag{9.5}$$

where v_{av} is the average cutting speed, mm/min; and D_{av} is average diameter of the work, mm.

The average diameter of the work is given by

$$D_{av} = (D_o + D_f)/2 \tag{9.6}$$

where D_f is the final diameter of the work-work, mm.

TABLE 9.2

Cutting Speed and Feed Data for *Turning* of Various Materials

Material	Alloy Steel (Normalized)	High Carbon Steel (annealed)	Mild Steel	Cast Iron	Bronze	Aluminum Alloys
Speed, sfm[a]						
Roughing	45	50	90	70	100	200
Finishing	60	65	120	80	130	300
Feed, in/rev						
Roughing	0.010–0.020	0.010–0.020	0.010–0.020	0.010–0.020	0.010–0.020	0.015–0.030
Finishing	0.003–0.005	0.003–0.005	0.003–0.005	0.003–0.010	0.003–0.010	0.005–0.010

[a]sfm = surface feet per minute.

The depth of cut is related to the initial and final diameters, as follows:

$$d = (D_o - D_f)/2 \qquad (9.7)$$

where d is the depth of cut, mm.

The feed rate is related to the feed and spindle rpm by

$$f_r = fN \qquad (9.8)$$

where f is feed, mm/rev; and f_r = feed rate, mm/min.

The cutting time can be computed by the following relationship:

$$T_c = L/(fN) \qquad (9.9)$$

where T_c is the cutting time, min; and L is the length of cut, mm.

The material removal rate (*MRR*) can be calculated by

$$MRR = v\,d\,f \qquad (9.10)$$

where *MRR* is the material removal rate, mm^3/min.

The number of cutting passes for machining can be determined by

$$\text{No. of cutting passes} = d/d' \qquad (9.11)$$

where d' is the depth of cut per pass.

In thread turning or threading, the pitch is given by (see Figure 9.4)

$$p = L/\text{No. of threads} \qquad (9.12)$$

where p is the pitch, mm, and L is length of cut, mm.

The number of revolutions required for threading can be determined by

$$\text{No. of rev} = L/p \qquad (9.13)$$

9.2.3 Machine Tool for Turning

The machine tool used for *turning* is called a *lathe*. There are a variety of lathes that are used for turning and related machining operations. *Engine lathe* is the most commonly used machine tool for turning. On an *engine lathe*, the tool is clamped onto a cross slide that is power driven on straight paths parallel or perpendicular to the work axis (see Figure 9.5).

9.3 Machining Operation—Drilling

General: Drilling produces a round hole in a workpiece by using a rotating tool (drill bit) with two cutting edges. The drilled hole may be either a *through hole*

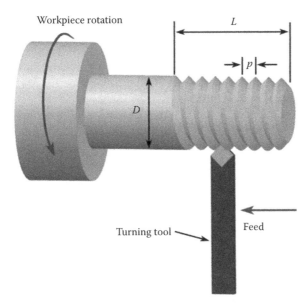

FIGURE 9.4
Thread turning—machining geometric parameters.

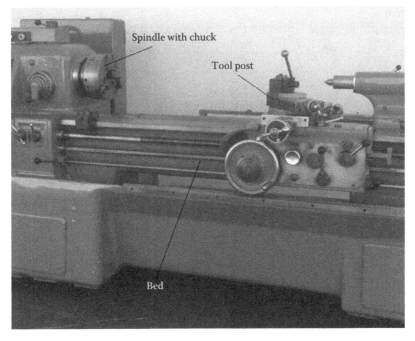

FIGURE 9.5
An *engine lathe*.

or a *blind hole* (see Figure 9.6). In a *through hole*, the hole exits the opposite side of the work, whereas in a *blind hole*, the hole does not exit the opposite side of the workpiece.

In addition to the (basic) drilling described in the preceding paragraph, there are other drilling-related operations; these include reaming, tapping, and the like. Reaming operation is used to slightly enlarge a hole (to remove all tool marks on the premachined hole); it is also used to achieve a better surface finish in the premachined hole. Tapping operation is used to provide internal screw threads on an existing hole.

Mathematical Modeling of Drilling: The cutting speed in drilling can be determined by

$$V = \pi\, D\, N \tag{9.14}$$

where V is the cutting speed, mm/min; D is the drill diameter, mm; and N is the spindle, rpm.

The cutting speed in drilling depends on the drill tool and the work material (see Table 9.3).

The material removal rate can be calculated by

$$MRR = (\pi\, D^2\, f_r)/4 \tag{9.15}$$

where MRR is the material removal rate, mm^3/min; and f_r is the feed rate, mm/min.

The cutting time in drilling a *through hole* can be computed by

$$T_c = L/f_r = (t + h)/f_r \tag{9.16}$$

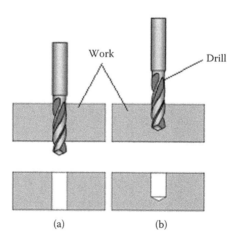

(a)   (b)

FIGURE 9.6
Two types of drilled hole: (a) *through hole* and (b) *blind hole*.

TABLE 9.3

The Cutting Speed in Drilling Using HSS Drill for Various Work Materials

Material	Al and Its Alloys	Bronze	Cast Iron (hard)	Carbon Steel	Tool Steel	Titanium Alloys
Cutting speed, sfpm	250	100	20	100	40	20

where T_c is the cutting time in drilling, min; L is the length of the cut, mm; t is the work thickness, mm; h is the drill-point height, mm.

The drill point height, h, can be calculated by the following formula:

$$h = (D/2)/[\tan (\phi/2)] \tag{9.17}$$

where D is the drill diameter, mm; and ϕ is the drill point angle ($\phi = 118°$).

The cutting time in drilling a *blind hole* can be computed by

$$T_c = d/f_r \tag{9.18}$$

where d is the hole depth, mm; and f_r is the feed rate, mm/min.

The torque, in turning, is related to the cutting force by

$$T = F_c D_{av}/2 \tag{9.19}$$

where T is the torque, N-m; F_c is the cutting force, N; and D_{av} is the average work diameter.

9.4. Problems and Solutions—Examples in Tool Life, Turning, and Drilling

EXAMPLE 9.1: COMPUTING THE CUTTING SPEED FOR A CUTTING TOOL MATERIAL WHEN TOOL LIFE IS KNOWN

Compute the cutting speed for steel cutting if a 100 minute tool life is required in each case: (a) HSS tool and (b) coated carbide tool.

Solution

$T_L = 100$ min, $v = ?$

(a) By reference to Table 9.1, for HSS, $n = 0.13$ (average), $C = 70$ m/min $= 70,000$ mm/min.

By using Equation 9.1,

$$v \, T_L^n = C$$

$$v\,(100)^{0.13} = 70{,}000$$

$$v = 38467.8\,\text{mm/min}$$

The cutting speed using HSS tool $= 38467.8\,\text{mm/min}$.
(b) By reference to Table 9.1, for coated carbide, $n = 0.5$, $C = 700\,\text{m/min} = 700{,}000\,\text{mm/min}$.

By using Equation 9.1,

$$v\,T_L^n = C$$

$$v\,(100)^{0.5} = 700{,}000$$

$$v = 70{,}000\,\text{mm/min}$$

The cutting speed using coated carbide cutting tool $= 70{,}000$ mm/min.

EXAMPLE 9.2: CALCULATING AND COMPARING *MRR* FOR TWO CUTTING TOOL MATERIALS FOR A GIVEN TOOL LIFE

In a turning operation of a steel workpiece, a feed of 0.3 mm/rev and depth of cut of 2 mm were used. Determine and compare *MRR* for each of the following tool materials if a 100 minute tool life is required in each case: (a) HSS and (b) coated carbide.

Solution

$f = 0.3\,\text{mm/rev}$, $d = 2\,\text{mm}$, $T_L = 100\,\text{min}$.

(a) By reference to the data in Example 9.1, for HSS tool, $v = 38467.8$ mm/min.

By using Equation 9.10,

$$MRR = v\,d\,f = 38467.8 \times 2 \times 0.3 = 23{,}081\,\text{mm}^3/\text{min}$$

For HSS tool, $MRR_1 = 23{,}081\,\text{mm}^3/\text{min}$.

(b) By reference to the data in Example 9.1, for coated carbide tool, $v = 70{,}000\,\text{mm/min}$.

By using Equation 9.10,

$$MRR = v\,d\,f = 70{,}000 \times 2 \times 0.3 = 42{,}000\,\text{mm}^3/\text{min}$$

For coated carbide tool, $MRR_2 = 42{,}000\,\text{mm}^3/\text{min}$.

By comparing the *MRR* for the two cutting tools,

$$(MRR_2)/(MRR_1) = 42{,}000/23{,}081 = 1.81$$

The use of coated carbide tool results in an increase of *MRR* by by almost double as compared to the HSS tool.

EXAMPLE 9.3: CALCULATING THE VALUES OF *n* AND *C* IN TAYLOR'S EQUATION USING MACHINING DATA

In a turning test using a coated carbide tool on hardened alloy steel, the tool failure occurred after 21 minutes when the cutting speed was 125 m/min. In another test using the same tool and work, the tool failed after 10 minutes when the cutting speed was 165 m/min. Calculate the values of *n* and *C* in Taylor's equation.

Solution

Case 1: $T_L = 21$ min, v = 125 m/min.

By using Equation 9.2,

$$\log 125 + n \log 21 = \log C$$

$$2.1 + n\,(1.3) = \log C \tag{E9.3.1}$$

Case 2: $T_L = 10$ min, v = 165 m/min.

By using Equation 9.2,

$$\log 165 + n \log 10 = \log C$$

$$2.22 + n = \log C \tag{E9.3.2}$$

By subtracting Equation (E9.3.2) from Equation E9.3.1, we get

$$-0.12 + 0.3\,n = 0$$

$$n = 0.4$$

By substituting $n = 0.4$ in Equation E9.3.2, we get

$$\log C = 2.22 + 0.4 = 2.62$$

$$C = 10^{2.62} = 417$$

Hence, $n = 0.4$, $C = 417$.

EXAMPLE 9.4: CALCULATING THE TOOL LIFE WHEN *n* AND *C* VALUES ARE KNOWN

By using the data in Example 9.3, calculate the tool life at a cutting speed of 100 m/min.

Solution

$n = 0.4$, $C = 417$, $v = 100$ m/min, $T_L = ?$

By using Equation 9.2,

$$\log v + n \log T_L = \log C$$

$$\log 100 + (0.4 \log T_L) = \log 417$$

$$0.4 \log T_L = \log (417/100) = \log 4.17 = 0.62$$

$$T_L = 10^{1.55} = 35.5 \text{ min.}$$

The cutting life $= 35.5$ min.

EXAMPLE 9.5: ASSESSING THE EFFECT OF % INCREASE IN CUTTING SPEED ON THE TOOL LIFE

For a tool wear, $n = 0.5$ and $C = 95$ in Taylor's equation. What is the percent increase in tool life if the cutting speed is reduced by 75%?

Solution

$V_2 = 75\%$ $V_1 = 0.75$ V_1.

By using Equation 9.1,

$$V_2 \, (T_2)^n = V_1 \, (T_1)^n$$

$$0.75 \, V_1 \, (T_2)^n = V_1 \, (T_1)^n$$

$$0.75 \, (T_2)^n = (T_1)^n$$

$$0.75 \, (T_2)^{0.5} = (T_1)^{0.5}$$

$$(T_2/T_1) = 1.33^2 = 1.78$$

$$T_2 = 1.78 \, T_1$$

$$\text{Increase in tool life} = T_2 - T_1 = 1.78 \, T_1 - T_1 = 0.78 \, T_1$$

$$\% \text{ Increase in tool life} = [(T_2 - T_1)/T_1] \times 100 = [(0.78 T_1)/T_1] \times 100 = 78$$

Increase in tool life $= 78\%$.

EXAMPLE 9.6: SELECTING AN APPROPRIATE CUTTING SPEED AND COMPUTING THE SPINDLE RPM

A cast iron cylindrical workpiece of length 800 mm and original diameter 120 mm is to be rough machined in an engine lathe to obtain 116 mm diameter work. (a) Select an appropriate maximum cutting speed and feed. (b) Compute the spindle rpm for the turning operation.

Solution

(a) By reference to Table 9.2, we get

$$v_{max} = 70 \text{ ft/min} = 21{,}336 \text{ mm/min}, f = 0.015 \text{ in/rev} = 0.38 \text{ mm/rev}$$

(b) By using Equation 9.4,

$$v_{max} = \pi D_o N$$

$$21{,}336 = \pi \times 120 \times N$$

$$N = 57 \text{ rpm}$$

The spindle rpm or work rotational speed $= 57$ rev/min.

EXAMPLE 9.7: CALCULATING THE CUTTING TIME FOR THE TURNING OPERATION

By using the data in Example 9.6, calculate the cutting time for the turning operation.

Solution

$L = 800$ mm, $f = 0.38$ mm/rev, $N = 57$ rpm.

By using Equation 9.9,

$$T = L/(f N) = (800)/(0.38 \times 57) = 37 \text{ min}$$

The cutting time $= 37$ minutes.

EXAMPLE 9.8: CALCULATING THE *MRR* FOR THE TURNING OPERATION

By using the data in Example 9.6, calculate the *MRR* for the turning operation.

Solution

$D_o = 120$ mm, $D_f = 116$ mm, $v_{max} = 21{,}336$ mm/min, $f = 0.38$ mm/rev.

By using Equation 9.7,

$$d = (D_o - D_f)/2 = (120 - 116)/2 = 2 \text{ mm}$$

By using Equation 9.10,

$$MRR = v \, d \, f = 21{,}336 \times 2 \times 0.38 = 16{,}215 \text{ mm}^3/\text{min}$$

Material removal rate $= 16{,}215$ mm^3/min.

EXAMPLE 9.9: COMPUTING CUTTING SPEED FOR THE TURNING OPERATION WHEN SPINDLE RPM IS UNKNOWN

A 500 mm long cylindrical workpiece, having a diameter of 130 mm, is to be turned in 6 min. If a feed of 0.35 mm/rev is used, what cutting speed must be used to perform the turning operation?

Solution

$L = 500$ mm, $D_o = 130$ mm, $f = 0.35$ mm/rev, $T = 6$ min, $v_{max} = ?$

By using Equation 9.9,

$$T = L/(f\,N)$$

$$6 = (500)/(0.35\,N)$$

$$N = 238 \text{ rpm}$$

By using Equation 9.4,

$$v_{max} = \pi\,D_o\,N = \pi \times 130 \times 238 = 97{,}240 \text{ mm/min} = 97.2 \text{ m/min}$$

The cutting speed $= 97.2$ m/min $= 1.62$ m/s.

EXAMPLE 9.10: DETERMINING THE NUMBER OF CUTTING PASSES AND SPINDLE RPM IN TURNING

A 240 mm long steel bar is to be machined (by straight turning) by using a feed of 0.12 mm/rev at a cutting speed of 30 m/min. The diameter of the bar is to be reduced from 140 mm to 132 mm. If the permissible depth of cut is 2 mm, determine (a) the number of cutting passes and (b) the work rotational speed (spindle rpm) for each pass.

Solution

$L = 240$ mm, $f = 0.12$ mm/rev, $v = 30$ m/min $= 30{,}000$ mm/min, $D_o = 140$ mm, $D_f = 132$ mm.

(a) Permissible depth of cut in a pass $= d' = 2$ mm.

By using Equation 9.7,

$$d = (D_o - D_f)/2 = (140 - 132)/2 = 8/2 = 4 \text{ mm}$$

By using Equation 9.11,

$$\text{Number of cutting passes} = d/d' = 4/2 = 2$$

Number of cutting passes $= 2$.

(b) For the first pass, $D_o = 140$ mm.
By using Equation 9.4,

$$v_{max} = \pi \, D_o \, N_1$$

$$30,000 = \pi \times 140 \, N$$

$$N_1 = 68.2 \, \text{rpm}$$

The work rotational speed in the first pass $= 68.2$ rpm.

For the second pass, $D_o =$ diameter after first pass $= 140 - 2d' = 140 - 4 = 136$ mm.

By using Equation 9.4,

$$v_{max} = \pi \, D_o \, N_2$$

$$30,000 = \pi \times 136 \, N$$

$$N_2 = 70.2 \, \text{rpm}$$

The work rotational speed in the second pass $= 70.2$ rpm.

EXAMPLE 9.11: CALCULATING THE TOTAL CUTTING TIME FOR A MULTIPLE PASSES TURNING OPERATION

By using the data in Example 9.10, calculate the total cutting time for the turning operation.

Solution

$L = 240$ mm, $f = 0.12$ mm/rev, $N_1 = 68.2$ rpm, $N_2 = 70.2$ rpm.

$$\text{Cutting time for the first pass} = T_1 = L/(f \, N_1) = (240)/(0.12 \times 68.2)$$
$$= 29.3 \, \text{min}$$

$$\text{Cutting time for the second pass} = T_2 = L/(f \, N_2) = (240)/(0.12 \times 70.2)$$
$$= 28.6 \, \text{min}$$

$$\text{Total cutting time} = T_1 + T_2 = 29.3 + 28.6 = 57.8 \, \text{min}.$$

EXAMPLE 9.12: COMPUTING THE NUMBER OF REVOLUTIONS REQUIRED IN THREADING

In a thread turning operation, a bolt is threaded along a length of cut of 50 mm with a pitch of 1.25 mm. How many revolutions of spindle are required? How many threads are cut?

Solution

$L = 50$ mm, $p = 1.25$ mm, Number of rev $= ?$

By using Equation 9.13,

$$\text{Number of rev} = L/p = 50/1.25 = 40$$

The number of revolutions required $= 40$.

By using Equation 9.12,

$$p = L/\text{Number of threads}$$

$$1.25 = 50/\text{Number of threads}$$

Number of threads $= 40$

This confirms that each revolution of workpiece results in a thread.

EXAMPLE 9.13: CALCULATING THE *MRR* IN DRILLING

A *through hole* is drilled in a metal plate by using a drill diameter of 35 mm. The cutting speed is 30 m/min and feed is 0.3 mm/rev. Calculate the *MRR*.

Solution

$D = 35$ mm, $V = 30$ m/min $= 30,000$ mm/min, $f = 0.3$ mm/rev, $MRR = ?$

By using Equation 9.14,

$$V = \pi D N$$

$$30,000 = 3.142 \times 35 \, N$$

$$N = 272.7 \text{ rpm}$$

By using Equation 9.8,

$$f_r = f N = 0.3 \times 272.7 = 81.8 \text{ mm/min}$$

$N = 272.2$ rpm, $f_r = 81.8$ mm/min, $D = 35$ mm.

By using Equation 9.15,

$$MRR = (\pi D^2 f_r)/4 = (3.142 \times 35^2 \times 81.8)/4 = 78{,}711 \text{ mm}^3/\text{min}$$

Metal removal rate $= 78{,}711 \text{ mm}^3/\text{min}$.

EXAMPLE 9.14: CALCULATING THE CUTTING TIME TO DRILL A THROUGH HOLE

By using the data in Example 9.13, calculate cutting time in drilling if work thickness is 40 mm.

Solution

$t = 40$ mm, $N = 272.2$ rpm, $f_r = 81.8$ mm/min, $D = 35$ mm, $\phi = 118°$.
By using Equation 9.17,

$$h = (D/2)/[\tan(\phi/2)] = (35/2)/[\tan(118/2)] = (17.5)/(\tan 59°) = 10.5 \text{ mm}$$

By using Equation 9.16,

$$T_c = L/f_r = (t + h)/f_r = (40 + 10.5)/81.1 = 0.62 \text{ min}$$

The cutting time in drilling $= 0.62$ min.

EXAMPLE 9.15: CALCULATING THE CUTTING TIME TO DRILL A BLIND HOLE

A blind hole is drilled at a depth of 45 mm in a metal workpiece by using a 25 mm diameter twist drill. The drilling cutting conditions include speed $= 22$ m/min and feed $= 0.20$ mm/rev. Determine the cutting time for the drilling operation.

Solution

$d = 45$ mm, $D = 25$ mm, $V = 22$ m/min $= 22,000$ mm/min. $f = 0.2$ mm/rev, $T_c = ?$
By using Equation 9.14,

$$V = \pi D N$$

$$22,000 = \pi \times 25 \, N$$

$$N = 22000/(\pi \times 25) = 280.2 \text{ rpm}$$

By using Equation 9.8,

$$f_r = f N = 0.2 \times 280.2 = 56 \text{ mm/min}$$

By using Equation 9.18,

$$T_c = d/f_r = 45/56 = 0.8 \text{ min}$$

The cutting time in drilling the blind hole $= 0.8$ minutes.

EXAMPLE 9.16: SELECTING THE CUTTING SPEED AND FEED WHEN THE TOOL AND WORK ARE SPECIFIED

A 1.0 in diameter hole is to be produced in a bronze plate by using a HSS drill. (a) Select the cutting speed and (b) calculate the spindle rpm.

Solution

(a) By reference to Table 9.3, the cutting speed for drilling in a bronze plate is 100 ft/min.
(b) $V = 100$ ft/min $= 1200$ in/min, $D = 1$ in.

By using Equation 9.14,

$$V = \pi D N$$

$$1200 = \pi \times 1 \times N$$

$$N = 1200/\pi = 382 \text{ rev/min}$$

Spindle rpm $= 382$ rev/min.

EXAMPLE 9.17: CALCULATING THE FEED RATE WHEN THE DRILL DIAMETER AND SPINDLE RPM ARE KNOWN

A twist drill manufacturer specifies that the feed for using a 1 in twist drill should be 0.012 in/rev. By using the data in Example 9.16, calculate the feed rate for the drilling operation.

Solution

$f = 0.012$ in/rev, $N = 382$ rev/min.

By using Equation 9.8,

$$f_r = f N = 0.012 \times 382 = 4.58 \text{ in/min}$$

The feed rate $= 4.58$ in/min $= 116.3$ mm/min.

EXAMPLE 9.18: CALCULATING THE CUTTING FORCE WITHOUT USE OF DYNAMOMETER

A 150 mm long, 12 mm diameter cylindrical brass was machined by turning. The final diameter is 10 mm. The torque is 3.5 N-m. Calculate the cutting force.

Solution

$L = 150$ mm, $D_o = 12$ mm, $D_f = 10$ mm, $T = 3.5$ N-m $= 3500$ N-mm, $F_c = ?$

By using Equation 9.6,

$$D_{av} = (D_o + D_f)/2 = (12 + 10)/2 = 11 \text{ mm}$$

By using Equation 9.19,

$$T = (F_c \, D_{av})/2$$

$$3500 = (F_c \times 11)/2$$

$$F_c = 636.4 \, \text{N}$$

The cutting force $= 636.4 \, \text{N}$.

Questions and Problems

9.1 Distinguish between the following terms with the aid of diagrams: (a) single point and multiple point cutting tools, (b) through hole and blind hole, and (c) straight turning and thread turning.

9.2 (a) What properties are required in a cutting tool material? (b) List at least five cutting tool materials in order of increasing performance.

9.3 What is meant by tool life? Write down the mathematical relations for the tool life.

9.4 The wear of a cutting tool follows the following relationship:

$$\log v = -0.35 \log T_L + 25$$

(a) What are the values of the constants n and C in Taylor's equation? (b) Write Taylor's equation using the n and C values.

9.5 (a) Briefly explain the turning operation with the aid of a sketch. (b) List the various turning-related operations.

9.6 Briefly describe the function of an engine lathe.

9.7 What are the various drilling-related operations? Define two of them.

P9.8 Calculate the cutting speed for steel cutting if a 130 minute tool life is required in each case: (a) cemented carbide tool and (b) CBN tool.

P9.9 In a turning operation of steel workpiece, a feed of 0.3 mm/rev and depth of cut of 2 mm were used. Determine and compare MRR for each of the following tool materials if a 130 min tool life is required in each case: (a) cemented carbide and (b) CBN tool.

P9.10 In a turning test, the tool failure occurred after 26 minutes when the cutting speed was 120 m/min. In another test using the same tool and work, the tool failed after 18 minutes when the cutting speed was 155 m/min. Calculate the values of n and C in Taylor's equation.

P9.11 For a tool wear, $n = 0.38$ and $C = 100$ in Taylor's equation. What is the percent increase in tool life if the cutting speed is reduced by 60%?

P9.12 An aluminum alloy cylindrical workpiece of length 700 mm and original diameter 110 mm is to be finish machined in an engine lathe to obtain 100 mm diameter work. (a) Select an appropriate maximum cutting speed and feed; (b) compute spindle rpm for the turning operation.

P9.13 By using the data in Example 9.6, calculate (a) the cutting time and (b) the MRR for the turning operation.

P9.14 A 200 mm long steel bar is to be machined (by straight turning) by using a feed of 0.13 mm/rev at a cutting speed of 28 m/min. The diameter of the bar is to be reduced from 148 mm to 140 mm. If the permissible depth of cut is 2 mm, determine (a) the number of cutting passes, (b) the work rotational speed (spindle rpm) for each pass, and (c) the total cutting time.

P9.15 A *through hole* is drilled in a 45 mm thick metal plate by using a drill diameter of 30 mm. The cutting speed is 20 m/min, and feed is 0.2 mm/rev. Calculate (a) the metal removal rate and (b) the cutting time.

P9.16 A 0.5 in diameter hole is to be produced in a carbon steel plate by using a HSS drill. (a) Select the cutting speed and (b) calculate the spindle rpm.

References

Drouillet, C., Karandikar, J., Nath, C., Journeaux, A-C., El Mansori, M., Kurfess, T. 2016. Tool life predictions in milling using spindle power with the neural network technique. *Journal of Manufacturing Processes*, 22, 161–168.

Huda, Z. 2017. *Materials Processing for Engineering Manufacture*. Zurich, Switzerland: Trans Tech.

Youssef, H.A., El-Hofy, H. 2008. *Machining Technology: Machine Tools and Operations*. Boca Raton, FL: CRC Press.

10

Machining III—Shaping, Milling, and Grinding

10.1 Shaping and Planing

Principles of Operation: In *shaping/planing*, a straight flat surface is produced by the use of a single-point cutting tool that is moved linearly relative to the workpiece. In *shaping*, the tool reciprocates with speed motion while the workpiece has feed motion (see Figure 10.1a). In *planing*, the workpiece reciprocates with speed motion while the tool has feed (see Figure 10.1b).

Equipment: In a *shaping machine*, the reciprocating speed motion of the cutting tool is provided by a *ram* that moves relative to a *column* (see Figure 10.2). The ram has a double-stroke motion; it moves faster on the return stroke than on the forward cutting stroke (Bradley, 1973).

Mathematical Modeling: The shaping operation can be mathematically modeled so as to develop mathematical relationships for calculating the average cutting speed (v_a), the number of double strokes to complete a job (n), the total machining time (T_c), and the material removal rate (MRR).

The average cutting speed can be calculated by

$$v_a = N\, L_s\, (1 + \text{k}) \qquad (10.1)$$

where v_a is the average cutting speed, mm/s; N is the number of double strokes/min; L_s is the stroke length, mm; and k is the return-cutting ratio; which is numerically defined as follows:

$$k = (\text{Return stroke time})/(\text{cutting stroke time}) = t_r/t_c \qquad (10.2)$$

The number of double strokes (n) to complete the job can be found by

$$n = w/f \qquad (10.3)$$

where w is the width of the work, mm; and f is the feed per double stroke, mm/stroke.

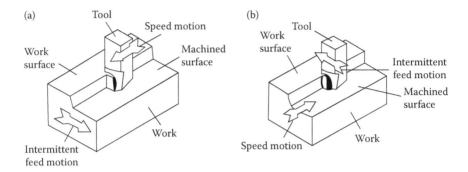

FIGURE 10.1
Schematic of shaping and planing operations: (a) shaping, (b) planing.

The total machining time is given by

$$T_c = n/N \tag{10.4}$$

where T_c is the machining time, min; n is the number of double strokes to complete the job; and N is the number of double strokes/min.

The MRR can be computed by the formula

$$MRR = d\,f\,v_a = d\,f\,N\,L_s\,(1+k) \tag{10.5}$$

where MRR is the metal removal rate, mm^3/min; and d is the depth of cut, mm.

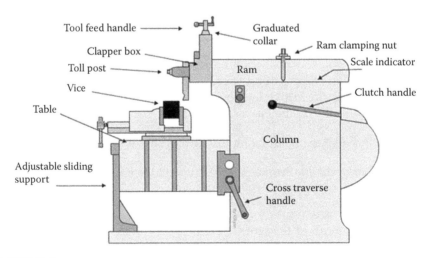

FIGURE 10.2
The parts of a shaper (shaping machine tool).

10.2 Milling

10.2.1 Principles of Operation in Milling

In *milling* a work is fed past a rotating cylindrical tool with multiple cutting edges. The *milling cutter* has *speed* motion, whereas the work has *feed* motion. There are two main forms of milling: (a) peripheral or slab milling and (b) face milling (Hall, 2004). In *peripheral milling*, the cutter axis is oriented parallel to the surface being machined, and the cutting edges are provided on the outside periphery of the cutter (see Figure 10.3). Recent machining studies have shown successful surface finishing of hardened steel by peripheral milling (Hayasaka et al., 2017). In *face milling*, the cutter axis is perpendicular to the surface being milled, and cutting edges are provided on both the end and outside periphery of the cutter (see Figure 10.4). There are two methods of milling: (1) up milling, and (2) down milling. In *up milling*, the cutter rotates in a direction opposite to the feed; this milling method is used for rough and deep cutting. In *down milling*, the cutter rotation is along the direction of the feed; this method is used to achieve a better surface finish.

10.2.2 Mathematical Modeling of Milling

The cutting speed in milling can be calculated by the formula

$$V = \pi D N \tag{10.6}$$

where V is the cutting speed, mm/min; D is the cutter outside diameter, mm; and N is spindle rpm. The feed rate can be calculated by the following mathematical relationship:

$$f_r = f_t N n_t \tag{10.7}$$

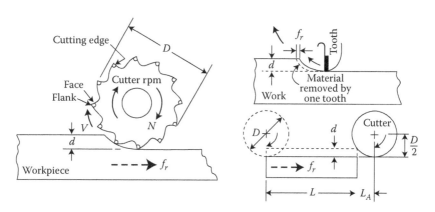

FIGURE 10.3
Geometric parameters in peripheral or slab milling.

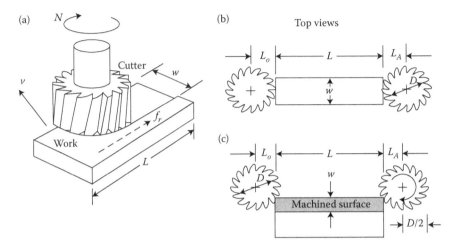

FIGURE 10.4
Geometric parameters in *face milling*: (a) face milling operation, (b) the cutter is centered over the work ($w > D/2$), and (c) partial face milling ($w < D/2$).

where f_r is the feed rate, mm/min; f_t is the feed/tooth, mm/tooth; and n_t is the number of teeth on the cutter (see Figure 10.3).

The MRR can be computed by

$$\text{MRR} = w \, d \, f_r \tag{10.8}$$

where MRR is the material removal rate, mm^3/min; w is the width of the cut; and d is the depth of the cut, mm.

The cutting time in *peripheral milling* can be found as follows:

$$T_c = (L + L_A)/f_r \tag{10.9}$$

where T_c is the cutting time, min; L is the length of the cut, mm; L_A is the approach length, mm (see Figure 10.3); and f_r is the feed rate, mm/min.

The approach distance (L_A) can be found as follows:

$$L_A = [d(D - d)]^{1/2} \tag{10.10}$$

The cutting time in *face milling* (Figure 10.4) can be calculated by the formula

$$T_c = (L + L_A + L_o)/f_r \tag{10.11}$$

where L is the length of the cut, mm; L_A is the approach length, mm; L_o is the length of "over travel," mm; and f_r is the feed rate, mm/min (see Figure 10.4).

It is evident in Figure 10.4 that the calculation for the approach length, L_A, depends on whether or not the cutter is centered over the workpiece.

When the cutter is centered over the work ($w > D/2$) (Figure 10.4b), the value of L_A is given as

$$L_A = L_o = D/2 \tag{10.12}$$

When the cutter is offset over the work ($w < D/2$) (Figure 10.4c), the value of L_A is given as

$$L_A = L_o = [w_c(D - w_c)]^{1/2} \tag{10.13}$$

where w_c is the width of swath cut by the cutter.

The torque in milling is given by

$$T = P_c/(2 \pi N) \tag{10.14}$$

where T is the torque, N-m; P_c is the cutting power, W; and N is spindle rotational speed, rev/s.

10.2.3 Milling Equipment

There are two types of milling machine: (a) horizontal milling machine and (b) vertical milling machine. In a *horizontal milling machine,* the main spindle is mounted horizontally. In a *vertical milling machine,* the spindle is mounted vertically in a head at the top of the column (see Figure 10.5).

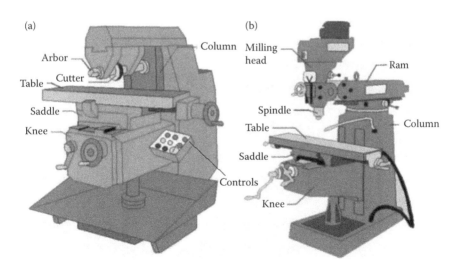

FIGURE 10.5
Milling machines: (a) horizontal, (b) vertical.

The milling operations on a *horizontal milling machine* include slotting, slitting, side milling, straddle milling, and milling of a gear. A *vertical milling machine* is capable of performing face milling, end milling, pocket milling, key milling, and profile milling.

10.2.4 Indexing in Milling and Its Mathematical Modeling

Indexing involves dividing a circular or straight part into equal spaces (e.g., for gear teeth cutting) by using an *indexing head* that is mounted on a milling machine (see Figure 10.6).

There are two types of indexing: (a) simple indexing and (b) differential indexing. In *simple indexing*, the index plate is fixed, whereas in *differential indexing* the index plate is rotated.

Direct or Simple Indexing: In simple indexing, the number of revolutions of index crank required for each division can be found by the formula

$$n = Z_o/Z \tag{10.15}$$

where n is the number of revolutions of index crank required for each division, Z_o is the number of teeth on a worm wheel ($Z_o = 40$), and Z is the number of divisions required on the work. The index plates with circles of holes patented by the *Brown and Sharpe Company* are shown in Table 10.1.

Differential Indexing: Differential indexing is employed where an index plate with the number of holes required for simple indexing is not available.

The following rules should be followed for differential indexing:

1. The number of divisions taken is Z', nearest to Z, which enables us to find on the available index plate a circular row with the number of holes allowing a division into Z' parts.

FIGURE 10.6
Indexing head mounted on a milling machine.

TABLE 10.1

The Index Plates Patented by the *Brown and Sharpe Company*

Plate Number	Number of Holes in the Index Plate
1	15, 16, 17, 18, 19, 20 holes
2	21, 23, 27, 29, 31, 33 holes
3	37, 39, 41, 43, 47, 49 holes

2. The number of revolutions of the index crank for simple indexing is found by $n = Z_o/Z'$.

3. The ratio of the differential change gears is determined by the following formula:

$$(Z_a/Z_b)\,(Z_c/Z_d) = [Z_o\,(Z' - Z)]/Z' \qquad (10.16)$$

4. If the ratio is positive, the index plate rotation is clockwise, and an idle gear is used.

5. If the ratio is negative, the index plate rotation is counterclockwise, and two idle gears must be used.

10.3 Grinding—Machining Operation

Grinding involves material removal by the action of hard and abrasive particles that are bonded usually in the form of a grinding wheel (Figure 10.7). It is generally used as (rough) finishing operations after part geometry has been established by conventional machining.

Grinding Wheel and Its Parameters: A grinding wheel is a disk consisting of abrasive grains and bonding material (see Figure 10.7). The abrasive grains act as cutting teeth during machining. Grinding wheels are precisely balanced for their high rotational speeds.

Grinding wheels are designated/specified by the following *wheel parameters* (Rowe, 2009):

1. Abrasive materials: Al_2O_3, SiC, CBN, diamond

2. Grit number: 8 (very coarse grains) to 600 (very fine grains)

3. Wheel grade: A to Z (soft to hard)

4. Wheel structure: 1 to 15 (1 = very dense and 15 = very open)

5. Bonding material: V, B, BF, S, R, RF, O, and M

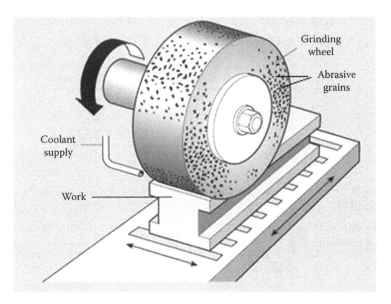

FIGURE 10.7
Surface grinding operation.

Grinding operations are usually performed to achieve a good surface finish. An excellent surface finish is achieved by (1) fine grain sizes (high grit number), (2) higher wheel speeds, and (3) denser wheel structure (more grits per wheel area).

Mathematical Modeling of Grinding: In grinding, the depth of cut (d) is much smaller and the cutting speed (v) is much higher as compared to other conventional machining operations.

The *surface speed* of the wheel is given by

$$v = \pi D N \tag{10.17}$$

where v is the *surface speed* of the grinding wheel, mm/min; D is the wheel diameter, mm; and N is the spindle rpm.

The length of a chip is given by

$$L_c = (D\,d)^{\frac{1}{2}} \tag{10.18}$$

where L_c is the length of the chip, mm; and d is the depth of cut, mm.

The *material removal rate* can be calculated by

$$\text{MRR} = v_w\, w\, d \tag{10.19}$$

where MRR is the material removal rate, mm^3/min; v_w is the work speed past the wheel, mm/min; and w is the width of the grinding path, mm.

10.4 Problems and Solutions—Examples in Shaping, Milling, and Grinding

10.4.1 Examples in Shaping Operation

EXAMPLE 10.1: COMPUTING THE RETURN-CUTTING RATIO AND THE CUTTING SPEED

The ram of a shaper takes 3 seconds during the cutting stroke and takes 2 seconds during the return stroke. The workpiece dimensions are 200 mm × 70 mm × 20 mm. The approach at each end is 60 mm. The ram makes 50 double strokes per minute. (a) What is the return-cutting ratio? (b) What is the stroke length? (c) Calculate the cutting speed for the shaping machine.

Solution

$t_c = 3$ s, $t_r = 2$ s, $L = 200$ mm, $L_A = 60$ mm, $N = 50$ double stroke/min, $k = ?$, $L_s = ?$, $v_a = ?$

(a) By using Equation 10.2,

$$k = t_r/t_c = 2/3 = 0.67$$

The return-cutting ratio $= k = 0.67$.

(b) The stroke length is the sum of the original length and two times the approach length.

$$L_s = L + 2\,L_A = 200 + (2 \times 60) = 320 \text{ mm}$$

The stroke length $= L_s = 320$ mm.

(c) By using Equation 10.1,

$$v_a = N\,L_s\,(1 + k) = 50 \times 320\,(1 + 0.67) = 26{,}720 \text{ mm/min} = 26.7 \text{ m/min}$$

The cutting speed of the shaper $= 26.7$ m/min.

EXAMPLE 10.2: COMPUTING THE TOTAL MACHINING TIME IN THE SHAPING OPERATION

By using the data in Example 10.1, compute the total machining time if a feed of 2 mm per double stroke is used in the shaping operation.

Solution

$w = 70$ mm, $N = 50$ double stroke/min, $f = 2$ mm/stroke, $T_c?$

By using Equation 10.3,

$$n = w/f = 70/2 = 35 \text{ mm}$$

By using Equation 10.4,

$$T_c = n/N = 35/50 = 0.7 \text{ min.}$$

Total machining time $= 0.7$ min.

EXAMPLE 10.3: COMPUTING THE MRR IN SHAPING MACHINING OPERATION

By using the data in Example 10.2, compute the MRR if the depth of the cut is 0.7 mm.

Solution

$d = 0.7$ mm, $f = 2$ mm/stroke, $v_a = 26{,}720$ mm/min, MRR $= ?$

By using Equation 10.5,

$$\text{MRR} = d\,f\,v_a = 0.7 \times 2 \times 26{,}720 = 37{,}408 \text{ mm}^3/\text{min}$$

Metal removal rate $= 37{,}408$ mm^3/min.

10.4.2 Examples in Milling

EXAMPLE 10.4: CALCULATING THE FEED RATE FOR THE MILLING OPERATION

It is required to remove 7.5 mm from the top surface of a metallic work with dimensions 270 mm × 80 mm × 15 mm by peripheral milling. The milling cutter has five teeth, and is 70 mm in diameter. The cutter overhangs the width of the work-part on both sides. The feed per tooth is 0.2 mm/tooth, and the cutting speed is 75 m/min. Calculate the feed rate for the milling operation.

Solution

$d = 7.5$ mm, $L = 270$ mm, $w = 80$ mm, $n_t = 5$, $D = 70$ mm, $f_t = 0.2$, $V = 75$ m/min $= 75{,}000$ mm/min.

By using Equation 10.6,

$$V = \pi\,D\,N$$

$$75{,}000 = \pi\,(70)\,N$$

$$N = 75{,}000/220 = 341 \text{ rpm}$$

By using Equation 10.7,

$$f_r = f_t\,N\,n_t = 0.2 \times 341 \times 5 = 341 \text{ mm/min}$$

The feed rate $= f_r = 341$ mm/min.

EXAMPLE 10.5: CALCULATING THE MACHINING TIME FOR ONE PASS BY PERIPHERAL MILLING

By using the data in Example 10.4, calculate the time to make one pass across the surface.

Solution

$L = 270$ mm, $D = 70$ mm, $d = 7.5$ mm, $f_r = 341$ mm/min.

In peripheral milling, the approach length is found by using Equation 10.10,

$$L_A = [d(D - d)]^{1/2} = [7.5 \, (70 - 7.5)]^{1/2} = 21.6 \, \text{mm}$$

By using Equation 10.9,

$$T_c = (L + L_A)/f_r = (270 + 21.6)/341 = 0.85 \, \text{min}$$

The time to make one pass across the surface $= 0.85$ min.

EXAMPLE 10.6: COMPUTING THE MRR IN PERIPHERAL MILLING

By using the data in Example 10.5, compute the MRR.

Solution

$d = 7.5$ mm, $f_r = 341$ mm/min, $w = 80$ mm.

By using Equation 10.8,

$$\text{MRR} = w \, d \, f_r = 80 \times 7.5 \times 341 = 204{,}600 \, \text{mm}^3/\text{min}$$

The material removal rate $= 204{,}600 \, \text{mm}^3/\text{min}$.

EXAMPLE 10.7: CALCULATING THE MACHINING TIME IN FACE MILLING WHEN $w > D/2$

By using the data for the workpiece, the cutter, and the cutting conditions in Example 10.4, calculate the time to make one pass across the surface, if the operation is face milling.

Solution

$L = 270$ mm, $w = 80$ mm, $f_r = 341$ mm/min (see solution of Example 10.4).

Here, $D/2 = 70/2 = 35$ mm, $w = 80$ mm, i.e., $w > D/2$; it means the cutter is centered over the work-part; hence, Equation 10.12 is applicable.

$$L_A = L_o = D/2 = 35 \, \text{mm}$$

By using Equation 10.11,

$$T_c = (L + L_A + L_o)/f_r = (270 + 35 + 35)/341 = 0.99 \text{ min} \sim 1 \text{ min}$$

The time to make one pass across the surface by face milling = 1 min.

EXAMPLE 10.8: CALCULATING THE MACHINING TIME IN FACE MILLING WHEN $w < D/2$

Repeat Example 10.7 except that the work is 75 mm wide and the cutter is offset to one side so that the swath cut by the cutter is 15 mm wide.

Solution

$L = 270$ mm, $D = 70$ mm, $f_r = 341$ mm/min, $w_c = 15$ mm, face milling with cutter offset.

$D/2 = 35$ mm, $w_c = 15$ mm; i.e., $w < D/2$; it means the cutter is offset to one side; hence, Equation 10.13 is applicable (see Figure 10.4c):

$$L_A = L_o = [w_c(D - w_c)]^{1/2} = [15(70–15)]^{1/2} = 28.7 \text{ mm}$$

By using Equation 10.11,

$$T_c = (L + L_A + L_o)/f_r = (270 + 28.7 + 28.7)/341 = 0.96 \text{ min}$$

Cutting time = 0.96 min.

EXAMPLE 10.9: CALCULATING THE MRR IN FACE MILLING WHEN $w < D/2$

By using the data in Example 10.8, calculate the MRR.

Solution

$d = 7.5$ mm, $f_r = 341$ mm/min, $w_c = 15$ mm.

By using Equation 10.8,

$$\text{MRR} = w \, d \, f_r = 15 \times 7.5 \times 341 = 38362.5 \text{ mm}^3/\text{min}$$

The MRR for the face milling when the cutter is offset = 38,362.5 mm³/min.

EXAMPLE 10.10: CALCULATING THE POWER AND TORQUE IN MILLING

A slab milling operation is performed on an annealed mild steel slab with dimensions: 320 mm × 110 mm. The depth of cut is 2.5 mm, and the feed per tooth is 0.27 mm/tooth. The milling cutter has an outer diameter of 55 mm, has 24 straight teeth, and rotates at 110 rev/min. The unit power for the work material is $3 \, \text{W} \cdot \text{s/mm}^3$. Calculate the following: (a) MRR, (b) power, and (c) torque.

Solution

$w = 110 \, \text{mm}$, $d = 2.5 \, \text{mm}$, $f_t = 0.27 \, \text{mm/tooth}$, $D = 55 \, \text{mm}$, $N = 110 \, \text{rpm}$, $n_t = 24$, $P_u = 3 \, \text{W} \cdot \text{s/mm}^3$.

(a) By using Equation 10.7,

$$f_r = f_t \, N \, n_t = 0.27 \times 110 \times 24 = 712.8 \, \text{mm/min}$$

By using Equation 10.8,

$$\text{MRR} = w \, d \, f_r = 110 \times 2.5 \times 712.8 = 196{,}020 \, \text{mm}^3/\text{min}$$

The metal removal rate $= \text{MRR} = 196{,}020 \, \text{mm}^3/\text{min} = 196{,}020/60 \, \text{s}$
$= 3267 \, \text{mm}^3/\text{s}$.

(b) By using Equation 8.15,

$$P_c = (P_u)(\text{MRR}) = 3 \times 3267 = 9801 \, \text{W}$$

Power $= 9801 \, \text{W}$.

(c) $P_c = 9801 \, \text{W}$, $N = 110 \, \text{rev/min} = 110/60 \, \text{rev/s} = 1.83 \, \text{rev/s}$.

By using Equation 10.14,

$$T = P_c/(2\pi N) = 9801/(2\pi \times 1.83) = 9801/11.5 = 852.2 \, \text{N-m}$$

Torque $= 852.2 \, \text{N-m}$.

EXAMPLE 10.11: CALCULATING NUMBER OF REVOLUTIONS FOR THE INDEX CRANK BY DIRECT INDEXING

Calculate the number of revolutions required for the index crank by direct indexing for dividing work in (a) 5 divisions, (b) 8 divisions, and (c) 32 divisions (identify the company and its plate number).

Solution

(a) $Z = 5$, $Z_o = 40$, $n = ?$

By using Equation 10.15,

$$n = Z_o/Z = 40/5 = 8$$

Hence, the index crank must make eight revolutions for each division on the work.

(b) $Z = 8$, $Z_o = 40$, $n = ?$

$$n = Z_o/Z = 40/8 = 5$$

Hence, the index crank must make five revolutions for each division on the work.

(c) $Z = 32$, $Z_o = 40$, $n = ?$

By using Equation 10.15,

$$n = Z_o/Z = 40/32 = 10/8 = 1\frac{2}{8}$$

$$\frac{2}{8} = \frac{2}{8} \times \frac{2}{2} = \frac{4}{16}$$

Thus, in this direct indexing, for each division on the job (work) the crank will move through one revolution and four holes on the 16-holes index circle on the index plate by using Plate 1 of the *Brown and Sharpe Company* (see Table 10.1).

EXAMPLE 10.12: CALCULATING THE NUMBER OF REVOLUTIONS FOR THE INDEX CRANK FOR BOLT HEAD

Calculate the number of revolutions of the index crank to machine a heptagonal head of a bolt using a milling machine. Identify the plates and the company.

Solution

$Z = 7$, $Z_o = 40$, $n = ?$

By using Equation 10.15,

$$n = Z_o/Z = 40/7 = 5\frac{5}{7}$$

$$\frac{5}{7} = \frac{5}{7} \times \frac{3}{3} = \frac{15}{21}$$

Hence, the index crank must make five revolutions and must move 15 holes on the 21-holes circle on *Plate 2* of the *Brown and Sharpe Company* (see Table 10.1).

EXAMPLE 10.13: DIFFERENTIAL INDEXING FOR GEAR

Select the index plate, and determine the number of revolutions of the index crank for cutting a gear with 53 teeth. Also select the differential change gears.

The available change gears are 24, 24, 28, 32, 36, 40, 44, 48, 56, 64, 72.

Solution

$Z = 53$, $Z' = 56$ (see Rule 1 of differential indexing); $Z_o = 40$.

By applying Rule 2 of differential indexing, $n = Z_o/Z'$,

$$n = \frac{40}{56} = \frac{5}{7} = \frac{5}{7} \times \frac{3}{3} = \frac{15}{21}$$

The index plate chosen is *Plate 2* (*Brown and Sharpe Company*) that has a circular row with 21 holes (see Table 10.1), and the sector fingers of the adjustable sector are set for 15 holes.

By using Equation 10.16,

$$(Z_a/Z_b)(Z_c/Z_d) = [Z_o(Z' - Z)]/Z' = [40(56 - 53)]/56$$

$$= \frac{40}{56} \times \frac{3}{1} = \frac{40}{56} \times \frac{72}{24}$$

Hence, the differential change gears are $Z_a = 40$ teeth, $Z_b = 56$ teeth, $Z_c = 72$, and $Z_d = 24$ teeth.

Here, $(Z' - Z) = 56 - 53 = +3$, i.e., a positive value, and simple gearing is to be used, the index plate rotation is clockwise, and an idler gear must be used (see Rule 4 of differential indexing).

10.4.3 Examples in Grinding Machining Operation

EXAMPLE 10.14: CALCULATING THE SPINDLE RPM AND THE LENGTH OF A CHIP IN GRINDING

In a surface grinding operation, the surface speed of the wheel is 1500 m/min, and the wheel diameter is 140 mm. The work speed past the wheel is 0.3 m/s and the depth of cut is 0.06 mm. The width of the grinding path is 6 mm. Calculate the following: (a) the spindle rpm and (b) the average length of a chip.

Solutions

$v = 1500$ m/min $= 1,500,000$ mm/min, $D = 140$ mm, $v_w = 0.3$ m/s $= 18$ m/min $= 18,000$ mm/min,

$w = 6$ mm, $d = 0.06$ mm

(a) By using Equation 10.17,

$$v = \pi D N$$

$$1,500,000 = \pi \times 140 N$$

$$N = (15,00,000)/(\pi \times 140) = (15,00,000)/(439.8) = 3411 \text{ rpm}$$

The spindle rotational speed $= 3411$ rpm.

(b) By using Equation 10.18,

$$L_c = (D d)^{\frac{1}{2}} = (140 \times 0.06)^{\frac{1}{2}} = 2.9 \text{ mm}$$

The average length of a chip $= 2.9$ mm.

EXAMPLE 10.15: CALCULATING THE MRR IN GRINDING

By using the data in Example 10.12, compute the MRR for the operation.

Solution

$v_w = 18,000$ mm/min; $w = 6$ mm, $d = 0.06$ mm, MRR = ?

By using Equation 10.19,

$$\text{MRR} = v_w\, w\, d = 18,000 \times 6 \times 0.06 = 6480 \text{ mm}^3/\text{min}$$

The MRR for the grinding operation $= 6480$ mm^3/min.

Questions and Problems

10.1 Differentiate between the following terms with the aid of diagrams: (a) shaping and planing and (b) peripheral milling and face milling.

10.2 Draw a labeled sketch of a shaping machine tool.

10.3 Distinguish between the following milling techniques: (a) up milling and down milling and (b) horizontal and vertical milling machines.

10.4 Briefly explain the grinding machining operation with the aid of a sketch.

10.5 What are the grinding cutting conditions for achieving an excellent surface finish?

P10.6 A face milling operation is used to machine 5 mm from the top surface of a rectangular piece of aluminum with dimensions: 400 mm × 100 mm. The cutter has five teeth (cemented carbide inserts) and is 130 mm in diameter. The cutting speed is 2.8 m/s, and the feed per tooth is 0.25 mm/tooth. Calculate the following: (a) time to make one pass across the surface and (b) MRR.

P10.7 It is required to remove 4 mm from the top surface of a slab of mild steel with dimensions: 330 mm × 90 mm × 8 mm by peripheral milling. A helical tooth plain milling cutter having 80 mm diameter and four teeth is used. The cutting speed and feed are 30 m/min and 0.3 mm/tooth, respectively. Calculate the machining time.

P10.8 Calculate the number of revolutions of the index crank to machine a square head of a bolt.

P10.9 It is required to machine a gear with 48 teeth using the simple indexing of the milling machine. Select the index plate, and calculate the number of turns of the index crank.

References

Bradley, I. 1973. *The Shaping Machine*. Madison, WI: Model and Allied Publications.

Hall, H. 2004. *Milling: A Complete Course (Workshop Practice)*. Philadelphia, PA: Trans-Atlantic.

Hayasaka, T., Ito, A., Shamoto, E. 2017. Generalized design method of highly-varied-helix end mills for suppression of regenerative chatter in peripheral milling. *Precision Engineering*, 48, April 2017, 45–59.

Rowe, W.B. 2009. *Principles of Modern Grinding Technology*. Waltham, MA: Elsevier.

11

Finishing/Surface Engineering

11.1 Abrasive Finishing Processes

In the preceding chapter, an account of grinding process was given; this chapter introduces finish grinding processes or abrasive finishing processes. Abrasive finishing processes involve the use of abrasive to achieve excellent super-finish (smooth surface). Examples of abrasive finishing operations include honing, lapping, super-finishing, polishing, and buffing. These operations are capable of producing mirror-like finishes with surface roughness values in the range from 0.01 to 0.025 μm. The *super-finishing* process is explained in the following paragraph, whereas the other abrasive finishing processes are discussed elsewhere (Marinescu et al., 2006).

Super-finishing involves using a reciprocating single-bonded abrasive stone against a rotating workpiece surface (see Figure 11.1). In super-finishing, a cutting fluid is used to cool the work surface and wash away chips. The cutting fluid builds up a lubricant film between the tool and work surface, which is why this process is capable of improving only the surface finish with no change in dimensions. A highly polished and smooth surface with surface roughness as low as 0.025 μm can be achieved by the super-finishing operation. Due to the increasing demand on many transmission parts (e.g., gears, crankshafts, camshafts), surface integrity along with appropriate fatigue performance is becoming a key issue in the automotive industry (Courbon et al., 2016).

11.2 Surface Hardening Processes

11.2.1 Basic Principle of Surface Hardening

Surface hardening or *case hardening* is generally applied to steels; it involves hardening the steel surface to a required depth without hardening the core of the component. A distinct advantage of surface hardening of steel is that the outer skin (case) is hard, whereas the inner core is left untouched so that the material still possesses the toughness/ductility properties (see Figure 11.2).

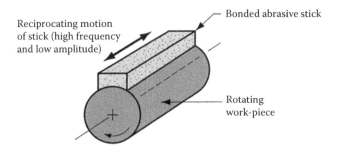

FIGURE 11.1
Super-finishing operation.

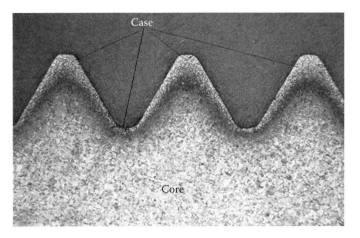

FIGURE 11.2
Surface hardened component (gear).

The surface hardening of steel has wide industrial applications; these include case-hardened gears, case-hardened grinding balls, and the like (Davis, 2003). There are two groups of surface hardening techniques: (1) surface hardening keeping the composition unchanged, and (2) surface hardening involving composition change; these techniques are discussed as follows.

11.2.2 Surface Hardening Keeping Composition the Same

There are two surface hardening processes that allow us to keep the composition unchanged: flame hardening and induction hardening.

Flame hardening: *Flame hardening* is a localized surface hardening heat-treatment process to harden the surface of steel containing moderate carbon contents. In *flame hardening*, a high-intensity oxy-acetylene flame is applied to the selective region so as to raise the surface temperature high enough in

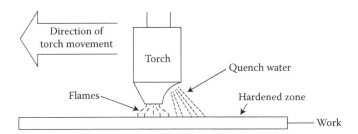

FIGURE 11.3
Flame hardening process.

the region of austenite transformation (see Figure 11.3). The overall heat transfer is restricted by the torch and thus the core (interior) never reaches the high temperature. The heated region (case) is quenched to achieve the desired hardness. Tempering can be done to eliminate brittleness, if desired.

Induction Hardening: Induction hardening involves the use of electric inductor coils to locally heat the steel component followed by fast cooling (quenching). The high-frequency electric fields rapidly heat the surface of the component, which is then quenched using water. This surface treatment results in a localized hardened layer at the surface.

11.2.3 Surface Hardening Involving Change in Composition

Principle: Steels may be surface hardened involving a change in composition by any of the following methods: (1) *carburizing*, (2) *nitriding*, and (3) *carbonitriding*. Surface hardening involving a change in composition is a diffusion-controlled process in which atoms of carbon, nitrogen, or both carbon and nitrogen are added to the surface by diffusion. For example, in carburizing, carbon atoms are added to the surface by a diffusion-controlled mechanism that is governed by Fick's law of diffusion.

Carburizing and Its Mathematical Modeling: There are two methods that are generally practiced for carburizing steels: gas carburizing and pack carburizing. In *gas carburizing*, steel is heated at a temperature in the range of 870–950°C in an atmosphere of carbon-rich gases, and then quenched. The equipment setup for *gas carburizing* is illustrated in Figure 11.4.

The diffusion of carbon during carburizing is based on the differential of concentration principle, which is expressed by Fick's second law of diffusion, as follows:

$$dC/dt = D(d^2C/dx^2) \tag{11.1}$$

where C is the % concentration of the diffusive atoms at a distance x from the surface; (dC/dt) is the rate of change of concentration; and D is diffusivity constant, m^2/s (see Figure 11.5).

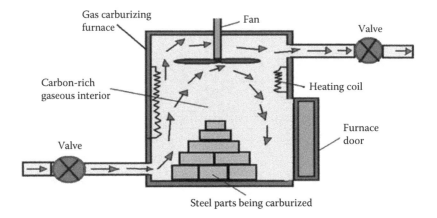

FIGURE 11.4
The equipment setup for *gas carburizing of steel.*

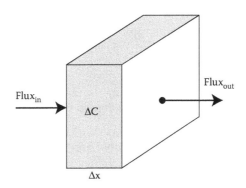

FIGURE 11.5
Fick's second law of diffusion.

On integrating both sides of Equation 11.1, we obtain

$$C = C_s - (C_s - C_o) \, erf \, \{x/[2(Dt)^{1/2}]\} \tag{11.2}$$

where C_o is the % initial bulk concentration, C_s is the % surface concentration, and *erf* is the Gaussian error function. A partial list of error function (*erf*) values is given in Table 11.1.

In case of approximate value of depth of case *x*, Equation 11.2 can be simplified to

$$C = C_s - (C_s - C_o) \, \{x/[2 \, (D \cdot t)^{1/2}]\} \tag{11.3}$$

where the terms have their usual meanings.

TABLE 11.1

Partial Listing of Error Function Values

z	erf (z)	z	erf (z)	z	erf (z)
0	0	0.30	0.3286	0.65	0.6420
0.025	0.0282	0.35	0.3794	0.70	0.6778
0.05	0.0564	0.40	0.4284	0.75	0.7112
0.10	0.1125	0.45	0.4755	0.80	0.7421
0.15	0.1680	0.50	0.5205	0.85	0.7707
0.20	0.2227	0.55	0.5633	0.90	0.7970
0.25	0.2763	0.60	0.6039	0.95	0.8209

In some metallurgical situations, it is desired to achieve a specific concentration of solute in an alloy; this situation enables us to rewrite Equation 11.2 as (Callister, 2007)

$$x^2/(D \cdot t) = \text{constant} \tag{11.4}$$

or

$$(x_1)^2/(D_1 \cdot t_1) = (x_2)^2/(D_2 \cdot t_2) \tag{11.5}$$

where x_1 is the case depth after carburizing for time duration of t_1; and x_2 is the case depth after carburizing for time duration of t_2. If the materials are identical and their carburizing temperatures are the same, the diffusivity constant is the same (i.e., $D_1 = D_2 = D$). Thus, Equation 11.5 simplifies to

$$(x_1)^2/(D \cdot t_1) = (x_2)^2/(D \cdot t_2)$$

or

$$x_1^2/t_1 = x_2^2/t_2 \tag{11.6}$$

It has been recently reported that the depth of the case in gas carburizing of steel exhibits a time–temperature dependence such that (Schneider and Chatterjee, 2013)

$$\text{Case depth} = Dt^{1/2} \tag{11.7}$$

Nitriding: This is the process of diffusion enrichment of the surface layer of a component with nitrogen. In *gas nitriding*, an alloy steel part is heated at a temperature in the range of 500°C–600°C for a duration in the range of 40–100 hours in the atmosphere of ammonia (NH_3), which dissociates to hydrogen and nitrogen. As a result, the nitrogen diffuses into the steel forming nitrides of iron, aluminum, chromium, and vanadium, thereby forming a thin hard surface.

Carbonitriding: This involves diffusion enrichment of the surface layer of a part with carbon and nitrogen. *Carbonitriding* is widely applied to the surface treatment of gears and shafts and is capable of producing a case depth in the range of 0.07–0.5 mm having hardness of 55–62 HR_C.

11.3 Surface Coating—An Overview

The surface coating process involves cleaning/surface preparation of a part (called substrate) followed by the application of a coating material; finally, the coated part is dried either in air or in an oven. Surface coatings provide protection, durability, and/or decoration to part surfaces. Metals that are commonly used as coating material (for coating on steels) include nickel, copper, chromium, gold, and silver. A good corrosion-resistant surface is produced by chrome plating; however, their durability is just fair. The nickel electroplated parts have excellent surface finish and brightness; they have a durable coating and have fair corrosion resistance.

There are many different surface coating processes; these include electroplating (nickel plating, chrome plating, etc.); hot dip coating, electroless plating, conversion coating, physical vapor deposition (PVD), chemical vapor deposition (CVD), powder coating, painting, and the like. Among these processes, electroplating is the most commonly practiced surface coating process and is discussed and mathematically modeled in the following section; the other coating processes are described elsewhere (Hughes et al., 2016; Huda, 2017).

11.4 Electroplating and Its Mathematical Modeling

11.4.1 Electroplating Principles

The electroplating process involves the cathodic deposition (plating) of a thin layer of metal on another metal (base metal) or other electrically conductive material. The electroplating process requires the use of direct current (*d.c.*) as well as an electrolytic solution consisting of certain chemical compounds that make the solution highly conductive. The positively charged plating metal (anode) ions in the electrolytic solution are drawn out of the solution to coat the negatively charged conductive part surface (cathode). Under an electromotive force (*e.m.f.*), the positively charged metal ions in the solution gain electrons at the part surface and transform into a metal coating (see Figure 11.6).

In modern surface engineering practice, electrolytic solutions usually contain additives to brighten or enhance the uniformity of the plating metal. The amount of plating material deposited strongly depends on the plating time and current levels to deposit a coating of a given thickness. The electroplating process is governed by Faraday's laws of electrolysis.

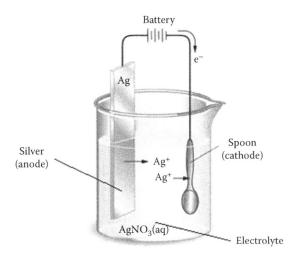

FIGURE 11.6
Electroplating process (silver plating of a steel spoon).

11.4.2 Mathematical Modeling of Electroplating Process

Mass of Metal Deposited: The mass of plated metal can be computed by the formula:

$$W = (ItA)/nF \qquad (11.8)$$

where W is the mass of the plated metal, g; I is the current, amperes; t is time, s; A is the atomic weight of the metal, g/mol; n is the valence of the dissolved metal in electrolytic solution in equivalents per mole; and F is the Faraday's constant ($=96,485$ Coul/equiv.).

Thickness of Deposit: The *thickness* of the plated metal, T, can be determined as follows:

$$T = W/(\rho \cdot S) \qquad (11.9)$$

where T is the thickness of the metal plated, cm; W is the mass of the plated metal, g; ρ is the density of the plating metal, g/cm³; and S is the surface area of the plated workpiece, cm².

Current Density: In electroplating industrial practice, *current density* is more important than current. A good quality coating requires smaller *current densities* but it take more electroplating time. The *current density* is related to *current* by

$$J = I/S \qquad (11.10)$$

where J is current density, A/cm²; I is current, A; and S is surface area of plated workpiece, cm².

Electroplating Time: The determination of *electroplating time* is also of great industrial importance since it has a direct impact on direct labor cost. A useful mathematical modeling can be developed by rearranging the terms in Equation 11.8 as follows:

$$t = (W \cdot n \cdot F)/(I \cdot A) \qquad (11.11)$$

where t is the electroplating time, s; W is the mass of the plated metal, g; I is the current, amperes; A is the atomic weight of the metal, g/mol; n is the valence of the dissolved metal in electrolytic solution in equivalents per mole; and $F = 96,500$ Coul/equiv.

Cathode Current Efficiency: It is defined as "the ratio of the weight of metal actually deposited to that, which would have resulted if all the current had been used for deposition." Numerically, *cathode current efficiency* is given by

$$Eff_{cathode} = W/W_{th} \qquad (11.12)$$

where $Eff_{cathode}$ is the cathode current efficiency; W is the weight of metal actually deposited, g; and W_{th} is the theoretical weight of the deposit, g. The theoretical weight can be found by

$$W_{th} = (Q \cdot A)/(F \cdot n) \qquad (11.13)$$

where Q is the electric charge flow, Coulomb or C; A is the atomic mass of the metal, g/mol; F is the Faraday's constant; and n is the valence of the metal deposit.

Electroplating Rate: It is the weight of the metal deposited per unit time. The electroplating rate can be determined by rearranging the terms in Equation 11.8 as follows:

$$R = W/t = (IA)/nF \qquad (11.14)$$

where R is the rate of electroplating, g/s; and other symbols have their usual meanings.

11.5 Problems and Solutions—Examples in Finishing/Surface Engineering

11.5.1 Examples in Surface Hardening

EXAMPLE 11.1: ESTIMATING THE DEPTH OF CASE IN DE-CARBURIZATION OF STEEL

An AISI-1017 boilerplate steel was exposed to air at a high temperature of 927°C due to malfunctioning of its heat exchanger. Calculate the approximate depth from the surface of the boiler plate at which the concentration of carbon decreases to one-half of its original content after exposure to the hot oxidizing environment for 20 hours. The concentration of carbon at the steel surface was 0.02%. The diffusivity constant of carbon in steel at 927°C is 1.28×10^{-11} m^2/s.

Solution

Diffusivity $= D = 1.28 \times 10^{-11}$ m^2/s; Time $= t = 20\,h = 72,000$ s.

Surface carbon concentration $= C_s = 0.02\%$.

Initial bulk carbon concentration $= C_o = 0.17\%$ (AISI-1017 steel contains 0.17% carbon).

Concentration of diffusive carbon atoms at distance x from the surface $= C = (1/2)\,(0.17\%) = 0.085\%$.

By using Equation 11.3,

$$C = C_s - (C_s - C_o)\,\{x/[2(Dt)^{1/2}]\}$$

$$0.085 = 0.02 - (0.02 - 0.17)\,\{x/[2\,(1.28 \times 10^{-11} \times 72,000)^{1/2}]\}$$

$$\text{or} \quad x = 3.03 \text{ mm}$$

Hence, the depth from the surface of the boiler plate at which the concentration of carbon decreases to one-half of its original content after exposure to the hot oxidizing environment for 20 hours, is 3 mm.

EXAMPLE 11.2: CALCULATING TIME REQUIRED TO CARBURIZE THE COMPONENT

A component made of 1015 steel is to be case-carburized at 1000°C. The component design requires a carbon content of 0.8% at the surface and a carbon content of 0.3% at a depth of 0.7 mm from the surface. The diffusivity of carbon in steel at 1000°C is 3.11×10^{-11} m^2/s. Compute the time required to carburize the component to achieve the design requirements.

Solution

Diffusivity $= D = 3.11 \times 10^{-11}$ m^2/s.

Surface carbon concentration $= C_s = 0.8\%$.

Initial bulk carbon concentration $= C_o = 0.15\%$. (The AISI-1015 steel contains 0.15% carbon)

Concentration of diffusive carbon atoms at distance x from the surface $= C = 0.3\%$; $x = 0.7$ mm.

By the application of Fick's second law of diffusion (Equation 11.2),

$$C = C_s - (C_s - C_o)\,erf\,\{x/[2\,(Dt)^{1/2}]\}$$

$$0.3 = 0.8 - (0.8 - 0.15)\,erf\,[0.0007/2\,(3.11 \times 10^{-11})^{1/2}\,t^{1/2})]$$

$$erf\,(62.7/\sqrt{t}) = 0.7692$$

Taking $Z = 62.7/\sqrt{t}$, we get $erf\,Z = 0.7692$

By reference to Table 11.1, we can develop a new table as follows:

erf Z	0.7421	0.7692	0.7707
Z	0.80	X	0.85

By using the interpolation mathematical technique,

$$(0.7692 - 0.7421)/(0.7707 - 0.7421) = (x - 0.80)/(0.85 - 0.80)$$

$$x - 0.8 = 0.047$$

$$x = 0.847 = Z$$

$$\text{or} \quad Z = 62.7/\sqrt{t} = 0.847$$

$$t = 5480\,s = 1.5\,h$$

The time required to carburize the component $= 1.5$ hours.

EXAMPLE 11.3: ESTIMATING TIME OF CARBURIZATION WHEN THE DIFFUSIVITY CONSTANT IS UNKNOWN

A carburizing heat treatment of a steel alloy for a duration of 11 hours raises the carbon concentration to 0.5% at a depth of 2.2 mm from the surface. Estimate the time required to achieve the same concentration at a depth of 4 mm from the surface for an identical steel at the same carburizing temperature.

Solution

$x_1 = 2.2$ mm when $t_1 = 11$ h; $x_2 = 4$ mm when $t_2 = ?$

By using Equation 11.6,

$$x_1^2/t_1 = x_2^2/t_2$$

$$(2.2)^2/11 = 4^2/t$$

$$t = 36.4\,h$$

The required carburization time $= 36.4$ hours.

11.5.2 Examples in Electroplating

EXAMPLE 11.4: COMPUTING THE MASS OF PLATING METAL IN ELECTROPLATING

A current of 1.5 ampere is passed for 45 minutes to a steel workpiece during a silver plating electrolytic process. Compute the mass of the silver metal that is plated on the steel workpiece.

Solution

$I = 1.5$ A; $t = 45$ min $= 2700$ s; $A = 107.8$ g/mol; $F = 96,485$ Coulombs/equivalent; $n = 1$; $W = ?$ g.

By using Equation 11.8,

$$W = (ItA)/nF = (1.5 \times 2700 \times 107.8)/(1 \times 96,485) = 4.5 \text{ g}$$

The mass of silver plated $= 4.5$ grams.

EXAMPLE 11.5: CALCULATING THE THICKNESS OF THE PLATING METAL IN ELECTROPLATING

A 1.2 gram of copper was electroplated on 25 cm^2 surface area of a steel workpiece. Calculate the thickness of the plated copper. The density of copper $= 8.9$ g/cm^3.

Solution

$S = 25$ cm^2, $W = 1.2$ g, $\rho = 8.9$ g/cm^3, $T = ?$

By using Equation 11.9,

$$T = W/(\rho \cdot S) = (1.2)/(8.9 \times 25) = 0.00539 \text{ cm}$$

The thickness of the plated copper $= 0.00539$ cm $= 53.9$ μm.

EXAMPLE 11.6: COMPUTING CURRENT WHEN CURRENT DENSITY AND WORK DIMENSIONS ARE GIVEN

The electroplating of copper on a cylindrical workpiece (length $= 15$ cm, $r = 5$ cm) was found to require a current density of 2 mA/cm^2. Calculate current for the electroplating process?

Solution

$r = 5$ cm, $h = 15$ cm, $J = 2$ mA/cm^2, $I = ?$

First we find the surface area of cylinder, S, as follows:

$$S = (2\pi r h) + (2\pi r^2) = (2\pi \times 5 \times 15) + (2\pi \times 5^2) = 471.2 + 157 = 628.3 \text{ cm}^2$$

By using Equation 11.10,

$$I = J \cdot S = 2 \times 628.3 = 1256.5 \text{ mA}$$

Current $= 1256.5 \times 10^{-3}$ A $= 1.25$ A.

EXAMPLE 11.7: CALCULATING THE ELECTROPLATING TIME

It is required to electroplate 27 grams of zinc on a steel workpiece by using a 15 amp current flow through a solution of $ZnSO_4$. Calculate the electroplating time.

Solution

$W = 27$ g, $I = 15$ amp, Atomic mass of $Zn = A = 65.4$ g/mol, $F = 96,500$ Coul/equiv., $t = ?$

The $ZnSO_4$ chemical formula indicates that zinc is divalent (Zn^{2+}) (i.e., $n = 2$).

By using Equation 11.11,

$$t = (W \cdot n \cdot F)/(I \cdot A = (27 \times 2 \times 96,500)/(15 \times 65.4) = 5312 \text{ s}$$

The electroplating time $= 5312 \text{ s} = 1.47 \text{ h}$.

EXAMPLE 11.8: CALCULATING THEORETICAL WEIGHT OF METAL DEPOSIT

A 10 amp current flows for 1.5 h for electroplating nickel on a steel workpiece. What is the *theoretical weight of the deposit*?

Solution

$I = 10$ A, $t = 1.5$ h $= 5400$ s, $A = 58.7$ g/mol; $n = 2$, $F = 96,500$ Coul/equiv.

$$\text{Electric Charge} = Q = I\,t = 10 \times 5400 = 54,000 \text{ C}$$

By using Equation 11.13,

$$W_{th} = (Q \cdot A)/(F \cdot n) = (54,000 \times 58.7)/(96,500 \times 2) = 16.4 \text{ g}$$

The theoretical weight of the nickel deposit $= 16.4$ g.

EXAMPLE 11.9: CALCULATING CATHODE CURRENT EFFICIENCY

By using the data in Example 11.8, calculate the *cathode current efficiency* if 15 g of nickel is actually deposited on the cathode.

Solution

$W = 15$ g, $W_{th} = 16.4$ g.

By using Equation 11.12,

$$\textit{Eff}_{cathode} = W/W_{th} = 15/16.4 = 0.91$$

$$\% \ \textit{Eff}_{cathode} = W/W_{th} \times 100 = 0.91 \times 100 = 91$$

Cathode current efficiency = 91%.

EXAMPLE 11.10: CALCULATING THE ELECTROPLATING RATE

By using the data in Example 11.8, calculate the electroplating rate for the process.

Solution

$I = 10$ amp, $A = 58.7$ g/mol; $n = 2$, $F = 96{,}500$ Coul/equiv., $R = ?$

By using Equation 11.14,

$$R = (IA)/(nF) = (10 \times 58.7)/(2 \times 96{,}500) = 0.003 \text{ g/s}$$

Electroplating rate $= 0.003 \times 3600$ g/h $= 11$ g/h.

Questions and Problems

11.1 What are the objectives of finishing/surface engineering processes?

11.2 (a) Define the term *surface hardening* with the aid of a diagram. (b) What is the distinct advantage of surface hardening? (c) What are the two groups of surface hardening processes?

11.3 Briefly explain the *flame hardening* process with the aid of a diagram.

11.4 (a) List the surface hardening processes that involve change in composition. (b) Explain the gas carburizing process with the aid of a diagram.

11.5 (a) Define *surface coating process*. (b) What are the objectives of surface coating? (c) Compare the merits and demerits of *chrome plating* and *nickel plating*. (d) List the various techniques of surface coating processes.

11.6 Explain *electroplating* with the aid of a diagram.

11.7 List the various abrasive finishing processes and explain one of them.

P11.8 A carburizing heat treatment of a steel alloy for a duration of 12 hours raises the carbon concentration to 0.55% at a depth of 2.3 mm from the surface. Estimate the time required to achieve the same concentration at a depth of 4.5 mm from the surface for an identical steel at the same carburizing temperature.

P11.9 A component made of 1010 steel is to be case-carburized at 927°C. The component design requires a carbon content of 0.6% at the surface and a carbon content of 0.25% at a depth of 0.4 mm from the surface. The diffusivity of carbon in steel at 927°C is 1.28×10^{-11} m^2/s.

Compute the time required to carburize the component to achieve the design requirements.

P11.10   A current of 2.5 amp is passed for 35 min to a steel workpiece during a nickel plating electrolytic process. Compute the mass of the nickel metal that is plated on the steel workpiece.

P11.11   Gold, 0.9 gram, was electroplated on 15 cm^2 surface area of a steel workpiece. Calculate the thickness of the plated copper. The density of gold $= 19.3 \text{ g/cm}^3$.

P11.12   A current of 4 amp was passed for electroplating of copper on a cylindrical workpiece (length $= 20$ cm, $r = 7$ cm). Calculate the current density used for the electroplating process.

References

Callister Jr., W.D., 2007. *Materials Science and Engineering: An Introduction*. New York: John Wiley and Sons.

Courbon, C., Valiorgue, F., Claudin, C., Jacquier, M., Dumont, F., Rech, J. 2016. Influence of Some Superfinishing Processes on Surface Integrity in Automotive Industry. *Procedia CIRP*, 45(2016), 99–102.

Davis, J.R. 2003. *Surface Hardening of Steels: Understanding the Basics*. Materials Park, OH: ASM International.

Huda, Z. 2017. *Materials Processing for Engineering Manufacture*. Zurich, Switzerland: Trans Tech.

Hughes, A.E., Mol, J.M.C., Zheludkevich, M.L., Buchheit, R.G. 2016. *Active Protective Coatings*. Dordrecht, Netherlands: Springer Materials Science Series (Publishers).

Marinescu, I.D., Uhlmann, E., Doi, T. 2006. *Handbook of Lapping and Polishing*. Boca Raton, FL: CRC Press/Taylor & Francis.

Schneider, M.J., Chatterjee, M.S. 2013. *Introduction to Surface Hardening of Steels*. Materials Park, OH: ASM International.

12

Powder Metallurgy

12.1 Steps in Powder Metallurgy

The powder metallurgy (P/M) process involves the following main steps: (1) metal powder production, (2) powder characterization, (3) mixing and blending, (4) compaction or pressing, and (5) sintering (see Figure 12.1) (Upadhyaya and Upadhyaya, 2011). The most commonly used metal powders are iron, copper, aluminum, tin, nickel, titanium, and refractory alloys. It is evident in Figure 12.1 that the first step in P/M is to produce metal powder of desired characteristics. The next step is to test or characterize the powder to determine its properties, size, shape, and other characteristics. Then the powder is mixed with various binders, lubricants, or other additives for facilitating further processing. In the next step, the powder mix/blend is compacted under pressure by use of a suitable press. Finally, the green compact is sintered (heated) in a controlled-atmosphere furnace.

12.2 Metal Powder Production and Its Mathematical Modeling

Principles and Practices: There are several methods by which metal powder may be produced; these techniques include (1) atomization of molten metal; (2) comminution of solid metal; (3) precipitation from solution of a salt; (4) thermal decomposition of a chemical compound; (5) reduction of a compound, usually the oxide, in the solid state, and (6) electrodeposition (Samal and Newkirk, 2015).

In the *atomization of molten metal method,* a stream of molten metal is broken up into small droplets by injecting pressured fluid in the form of jet (see Figure 12.2). These droplets are rapidly frozen before they come into contact with each other or with a solid surface (chamber bottom).

Mathematical Modeling: Inert-gas atomized metal powders solidify more rapidly in comparison to water atomized powders. The finer the particle size (the higher the surface-to-volume ratio), the higher is the *cooling rate*. The *cooling rate* of the metal droplets can be calculated as follows (Schulz, 2001):

$$dT/dt = [6\,h(T_m - T_g)]/(\rho \cdot c_{pm} \cdot d_m) \tag{12.1}$$

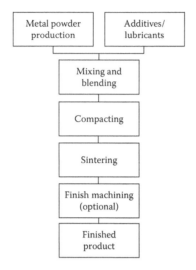

FIGURE 12.1
Steps in powder metallurgy (*P/M*).

where dT/dt is the cooling rate, K/s; h is the *convective heat transfer coefficient*, $J/s \cdot cm^2$; T_m is the metal-droplet temperature, K; T_g is the inert-gas temperature, K; ρ is the density of metal, g/cm^3; c_{pm} is the specific heat capacity of the metal, $J/g \cdot K$; and d_m is the droplet diameter, cm.

A rapid solidification processing by *spray atomization* is used to produce exceptionally fine secondary dendritic arm spacing (*SDAS*); a common method is to produce very fine liquid droplets that freeze into solid particles

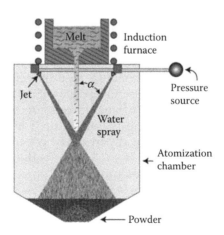

FIGURE 12.2
Metal powder production by atomization of molten metal.

having sizes in the range of 5–100 μm (Askeland and Wright, 2016). The *SDAS* can be computed by (Shiwen et al., 2009)

$$\lambda_2 = B(dT/dt)^{-n} \qquad (12.2)$$

where λ_2 is the *SDAS*, μm; dT/dt is the cooling rate, K/s; and B and n are constants having values of 47 and 0.33, respectively (see Example 12.2).

12.3 Metal Powder Characterization and Its Mathematical Modeling

A powder is a finely divided solid; each powder particle is smaller than 1 mm in its maximum. The processing of metal powders and the final properties of the sintered P/M products strongly depends on the characteristics of the powders. The powder characteristics include (1) particle shape, (2) particle size, (3) particle-size distribution, (4) chemical composition, (5) structure, (6) surface conditions, (7) powder flow rate, and (8) apparent density.

12.3.1 Particle Shape of Metal Powders

Metal powders may be in various particle shapes: spherical, spheroidal, angular, spongy, flakey, cylindrical, acicular, and cubic (see Figure 12.3).

Angular-shaped metal powders possess greater strength as compared to other shapes; however, their porosity is low. Spherical-shaped powders have greater flow properties (easier compaction) and porosity, but their strength is lower in comparison to angular-shaped metal powders.

The shape of cylindrical and acicular particles is determined by its aspect ratio, δ, which is defined as the ratio of its major dimension to its minor dimension. Numerically,

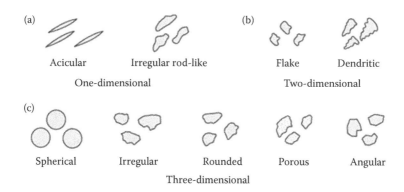

(a) Acicular — Irregular rod-like — One-dimensional

(b) Flake — Dendritic — Two-dimensional

(c) Spherical — Irregular — Rounded — Porous — Angular — Three-dimensional

FIGURE 12.3
Particle shapes of metal powders.

$$\delta = l/w \tag{12.3}$$

where l is the rod length and w is the width of the cylindrical or acicular particle.

The surface area (s) of a cylindrical or acicular particle is given by

$$s = (\pi w^2/2) + (\pi w l) \tag{12.4}$$

By combination of Equations 12.3 and 12.4,

$$\delta = (s/\pi w^2) - 0.5 \tag{12.5}$$

where δ is the aspect ratio; s is the surface area, mm^2; and w is the width of the particle, mm.

The specific surface area for a single particle is given by

$$\varepsilon = s/m \tag{12.6}$$

where ε is the specific surface area, mm^2/mg; and m is the mass of the particle (in mg) having the surface area s (in mm^2).

12.3.2 Particle Size of Metal Powders

Metal-powder particle size can be measured by use of a screen. The mesh count (MC) of a screen refers to the number of openings per linear inch of the screen (see Figure 12.4). The higher the MC, the smaller is the particle size. The length of a mesh in a screen is related to MC by

$$l_m = 1/(MC) \tag{12.7}$$

where l_m is the length of a mesh in the screen, and MC is the mesh count.

The MC of a powder particle can be calculated by

$$MC = \sqrt{N_O} \tag{12.8}$$

where MC is the mesh count, and N_O is total number of openings per square inch in the screen.

12.3.3 Particle Size Distribution of Metal Powders

The calculations for the particle size distribution of metal powders involve a great deal of statistical analysis. The arithmetic mean of the particle size distribution data is computed as follows. If a powder has particle sizes $x_1, x_2, x_3 \ldots$ with frequencies $f_1, f_2, f_3, \ldots$, the arithmetic mean (AM), quadratic mean (QM), and cubic mean (CM) for the particle size distribution are given by (Anwar, 2017):

$$AM = [\Sigma(x_i \cdot f_i)]/\Sigma f_i \tag{12.9}$$

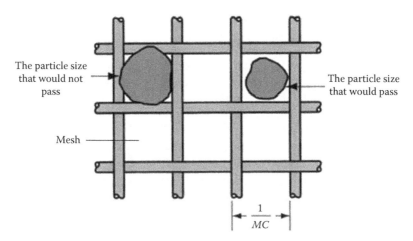

The particle size that would not pass

The particle size that would pass

Mesh

$$\frac{1}{MC}$$

FIGURE 12.4
Measurement of powder particle size by using a mesh screen.

$$QM = [\sum (x_i \cdot f_i)^2]^{1/2} / \sum f_i \qquad (12.10)$$

$$CM = [\sum (x_i \cdot f_i)^3]^{1/3} / \sum f_i \qquad (12.11)$$

The significance of Equations 12.9 through 12.11 is illustrated in Examples 12.6 through 12.8.

12.3.4 Flow Rate and Apparent Density of Metal Powder

The flow rate and apparent density of metal powders can be measured by using the *Hall flowmeter*.

In accordance with international norms, the time taken by 50 grams of powder to pass a standard funnel orifice (diameter = 2.5 cm) is determined.

In order to determine the apparent density of powders, a stainless steel density cup (25 cm^3) is provided according to international norm. First, the dry test specimen is loaded and filled into the flowmeter funnel permitted to run into the density cup through the discharge orifice. Then, the funnel is rotated 90° in a horizontal plane. The powder is then leveled off flush with the top of the density cup. Then, the density cup is lightly tapped on the side to settle the powder to avoid spilling in transfer. Finally, the powder is transferred to the balance and weighed.

The apparent density can be calculated by

$$\rho_a = 0.04\, w \qquad (12.12)$$

where ρ_a is the apparent density, g/cm^3; and w is the mass of powder, g.

12.4 Mixing and Blending and Its Mathematical Model

P/M components are usually manufactured from mixes of unalloyed or low-alloyed iron powder with additives like graphite, other metal powders, and lubricants. The density of the compact attainable with such powder mixes is strongly influenced by the densities and the relative amounts of the additives and of impurities (if any).

The theoretically achievable pore-free density of a powder mix can be calculated as follows:

$$\rho_M = 100/[(w_{Fe}/\rho_{Fe}) + (w_1/\rho_1) + (w_2/\rho_2) + \cdots] \qquad (12.13)$$

where ρ_M is the theoretically achievable pore-free density of a powder mix, g/cm^3; w_{Fe} is the weight percentage of iron powder; ρ_{Fe} is the density of iron powder; $w_1, w_2, \ldots$ are the weight% of additives and impurities, g; and $\rho_1, \rho_2, \rho_3, \ldots$ are the densities of additives and impurities.

12.5 Powder Compaction and Its Mathematical Model

Principle: Metal powders are generally compacted by using an *axial pressing* technique. In *axial pressing,* the metal powder is first added into a closed tool consisting of three main parts: a die, an upper punch, and a lower punch. Then the powder is subjected to a pressure by using a press (see Figure 12.5). Finally, the "green compact" is ejected.

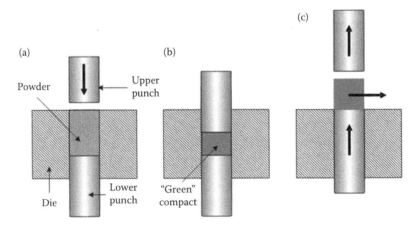

FIGURE 12.5
Stages in metal powder compaction; (a) filling powder in die cavity, (b) powder compaction, and (c) ejecting "green" compact.

Mathematical Model: Most applications for P/M dictate that high densities be obtained in the final product; this objective requires an external pressure, P. For a single-ended pressing, the average compaction stress, σ, is given by (Hofmann and Bowen, 2017):

$$\sigma = P[1 - \mu z(H/D)] \tag{12.14}$$

where μ is the coefficient of friction between the powder and the die wall; z is the constant that depends on compact density; H is the height of the cylindrical compact; and D is the diameter of the compact.

The required filling depth for the part can be calculated by means of the ratio Q between compact density and filling density (apparent density) of the powder according to the following relationship (Höganäs, 2013):

$$Q = \rho_c/\rho_a = d_f/H \tag{12.15}$$

where ρ_c is the compact density, g/cm^3; ρ_a is the filling density (apparent density) of the powder, g/cm^3; d_f is the depth of fill, cm; and H is the height of the cylindrical compact, cm.

12.6 Sintering of Compact

Sintering is the generally the final stage in P/M. It involves thermal treatment of a powder compact at a temperature below the melting point of the main constituent. *Sintering* results in an increase of the compact strength by bonding together of the particles. Thus, sintering is associated with shrinkage. The *percentage of shrinkage* can be calculated by

$$\% \text{ Shrinkage} = [(\Delta l)/l_s] \times 100 \tag{12.16}$$

where Δl is the change in length; and l_s is the sintered length.

12.7 Problems and Solutions—Examples in Powder Metallurgy

EXAMPLE 12.1: CALCULATING THE COOLING RATE OF METAL DROPLETS

Iron powder was produced by using an inert-gas atomization technique. The iron droplets temperature was 1600°C; the diameter of a droplet was measured to be 10 μm. The inert gas was injected into the atomization chamber at a temperature of 730°C. The convective heat transfer coefficient for the metal powder production system is 0.67 J/s cm^2. Calculate the *cooling rate* of the metal droplets.

Solution

By reference to the Physical Properties table,

Density of iron $= \rho = 7.8$ g/cm^3; Specific heat capacity of iron $= c_{pm} = 0.45$ J/g K.

$T_m = 1600°C = 1600 + 273 = 1873$ K $\quad T_g = 730°C = 1000$ K.

$d_m = 10\,\mu\mathrm{m} = 10 \times 10^{-6}$ m $= 10^{-5}$ m $= 10^{-3}$ cm $= 0.001$ cm; $h = 0.67$ J/s cm^2
$\mathrm{d}T/\mathrm{d}t = ?$

By using Equation 12.1,

$$\mathrm{d}T/\mathrm{d}t = [6h(T_m - T_g)]/(\rho \cdot c_{pm} \cdot d_m)$$

$$\mathrm{d}T/\mathrm{d}t = [6 \times 0.67(1873 - 1000)]/(7.8 \times 0.45 \times 0.001) = 10^6 \text{ K/s}$$

The cooling rate of the metal droplets $= 10^6$ K/s.

EXAMPLE 12.2: COMPUTING THE *SDAS*

By using the data in Example 12.1, calculate the *SDAS* for the ultrafine powder particles.

Solution

$\mathrm{d}T/\mathrm{d}t = 10^6$ K/s; $B = 47$; $n = 0.33$ (see Equation 12.2).

By using Equation 12.2,

$$\lambda_2 = B(\mathrm{d}T/\mathrm{d}t)^{-n} = 47(10^6)^{-0.33} = 47(10)^{-1.98} = 47/10^{1.98} = 47/95.5$$
$$= 0.49\,\mu\mathrm{m}$$

The *SDAS* $= 0.49\,\mu$m.

EXAMPLE 12.3: CALCULATING THE ASPECT RATIO AND SUR-FACE AREA

The rod length and width of an acicular metal-powder particle are 1.2 mm and 0.3 mm, respectively. Calculate its (a) aspect ratio and (b) surface area.

Solution

$l = 1.2$ mm, $w = 0.3$ mm, $\delta = ?$, $s = ?$

(a) By using Equation 12.3,

$$\delta = l/w = 1.2/0.3 = 4$$

The aspect ratio of the particle $= 4$.
(b) By using Equation 12.4,

$$s = (\pi w^2/2) + (\pi wl) = \{[\pi(0.3)^2]/2\} + (\pi \times 0.3 \times 1.2) = 1.27 \text{ mm}^2$$

The surface area of the particle $= 1.27$ mm^2.

EXAMPLE 12.4: CALCULATING THE SPECIFIC SURFACE AREA

By using the data in Example 12.3, calculate the specific surface area of the particle weighing 0.2 mgf.

Solution

$s = 1.27\,\text{mm}^2$; $m = 0.2\,\text{mg}$, $\varepsilon = ?$

By using Equation 12.6,

$$\varepsilon = s/m = 1.27/0.2 = 6.35\,\text{mm}^2/\text{mg}$$

The specific surface area of the particle $= 6.35\,\text{mm}^2/\text{mg}$.

EXAMPLE 12.5: CALCULATING THE MESH COUNT (*MC*) AND THE LENGTH OF MESH

There are 40,000 openings per square inch in a mesh screen. Calculate (a) the *MC* and (b) the length of a mesh in the screen.

Solution

Total number of openings per square inch in the screen $= N_O = 40,000$.

(a) By using Equation 12.8,

$$MC = \sqrt{N_O} = \sqrt{40,000} = 200$$

The mesh count $= MC = 200$.

(b) By using Equation 12.7,

$$l_m = 1/(MC) = 1/200 = 0.005\,\text{in.}$$

The length of a mesh in the screen $= 0.005\,\text{in.}$

EXAMPLE 12.6: ESTIMATING ARITHMETIC MEAN FOR PARTICLE SIZE DISTRIBUTION OF METAL POWDER

The particle size distribution data for a metal powder is given in Table E12.6a. Estimate the arithmetic mean.

Solution

By using the data in Table E12.6a, we can develop a (new) Table E12.6b as follow.

TABLE E12.6a

Particle Size Distribution for a Metal Powder

Particle size, μm	2.4	2.8	5.0	11.4	23.0	48.6	60.3	76.8	88.7
Cumulative % undersize	0.2	0.7	2.6	9.0	46.7	78.5	84.3	92.7	100

TABLE E12.6b

Particle Size Distribution Data for
Arithmetic Mean

x_i	f_i	%f	$x_i \cdot f_i$
2.4	0.2	0.2	0.48
2.8	0.5	0.7	1.4
5.0	1.9	2.6	9.5
11.4	6.4	9.0	73
23.0	37.7	46.7	867.1
48.6	31.8	78.5	1545.5
60.3	5.8	84.3	349.7
76.8	8.4	92.7	645.1
88.7	7.3	100	647.5
Σ 319	100		4139.3

By using Equation 12.9,

$$AM = [\textstyle\sum(x_i \cdot f_i)]/\sum f_i = 4139.3/100 = 41.4$$

The arithmetic mean of the particle size distribution $= 41.4\,\mu m$.

EXAMPLE 12.7: ESTIMATING QUADRATIC MEAN FOR PARTICLE SIZE DISTRIBUTION OF METAL POWDER

By using the data in Example 12.6, estimate the quadratic mean for the particle size distribution.

Solution

By using the data in Table E12.6a, we can develop a (new) Table E12.7 as follows.

By using Equation 12.10,

$$QM = \left[\textstyle\sum(x_i \cdot f_i)^2\right]^{1/2}/\sum f_i = (4103554)^{1/2}/100 = \frac{\sqrt{4103554}}{100} = 20.25$$

The quadratic mean of the particle size distribution $= 20.25\,\mu m$.

EXAMPLE 12.8: ESTIMATING CUBIC MEAN FOR PARTICLE SIZE DISTRIBUTION OF METAL POWDER

By using the data in Table E12.6a, estimate the cubic mean for the particle size distribution.

Solution

By using the data in Table E12.6a, we can develop a (new) Table E12.8 as follows.

By using Equation 12.11,

$$CM = [\sum(x_i \cdot f_i)^3]^{1/3}/\sum f_i = (4.9 \times 10^9)^{1/3}/100$$
$$= [(4.9)^{1/3} \times 1000]/100 = 36.6$$

The cubic mean of the particle size distribution $= 36.6\,\mu m$.

TABLE E12.7

Particle Size Distribution Data for Quadratic Mean

	x_i	f_i	$\%f$	$x_i \cdot f_i$	$(x_i \cdot f_i)^2$
	2.4	0.2	0.2	0.48	0.23
	2.8	0.5	0.7	1.4	1.96
	5.0	1.9	2.6	9.5	90.25
	11.4	6.4	9.0	73	5329
	23.0	37.7	46.7	867.1	751862.4
	48.6	31.8	78.5	1545.5	2388570.2
	60.3	5.8	84.3	349.7	122290.1
	76.8	8.4	92.7	645.1	416154
	88.7	7.3	100	647.5	419256.25
$\sum$	319	100		4139.3	4103554

TABLE E12.8

Particle Size Distribution Data for Cubic Mean

	x_i	f_i	$\%f$	$x_i \cdot f_i$	$(x_i \cdot f_i)^3$
	2.4	0.2	0.2	0.48	0.11
	2.8	0.5	0.7	1.4	2.74
	5.0	1.9	2.6	9.5	857.4
	11.4	6.4	9.0	73	389017
	23.0	37.7	46.7	867.1	651939895.7
	48.6	31.8	78.5	1545.5	3691535321.4
	60.3	5.8	84.3	349.7	42764844.4
	76.8	8.4	92.7	645.1	268460951.8
	88.7	7.3	100	647.5	271468421.8
$\sum$	319	100		4139.3	4.9×10^9

EXAMPLE 12.9: CALCULATING THE APPARENT DENSITY OF METAL POWDER

A metal powder was tested in a *Hall flowmeter* and then weighed; the mass of the powder was measured to be 12.41 g. Calculate the apparent density of the powder.

Solution

$w = 12.41$ g, $\rho_a = ?$

By using Equation 12.12

$$\rho_a = 0.04w = 0.04 \times 12.41 = 0.496$$

The apparent density $= 0.5$ g/cm^3.

EXAMPLE 12.10: CALCULATING THEORETICALLY ACHIEVABLE PORE-FREE DENSITY OF THE POWDER MIX

A 94 wt% iron powder was blended with 4 wt% graphite and 2 wt% impurity (with density 8 g/cm^3). Calculate the theoretically achievable pore-free density of the powder mix.

Solution

$w_{Fe} = 94$, $\rho_{Fe} = 7.8$ g/cm^3, $w_1 = 4$, $\rho_1 = 2.26$ g/cm^3, $w_2 = 2$, $\rho_1 = 8$ g/cm^3

By using Equation 12.13,

$$\rho_M = 100/[(w_{Fe}/\rho_{Fe}) + (w_1/\rho_1) + (w_2/\rho_2)]$$
$$= 100/[(94/7.8) + (4/2.26) + (2/8)] = 7.1 \text{ g/cm}^3$$

The theoretically achievable pore-free density of the powder mix $= 7.1$ g/cm^3.

EXAMPLE 12.11: CALCULATING THE COMPACTION STRESS IN METAL-POWDER PRESSING

For a single-action pressing of copper using a constant compaction pressure of 690 MPa, consider the following. The height-to-diameter ratio of a metal-powder compact is 0.8. The coefficient of friction between the powder and die wall is 0.3. Express the average compaction stress in terms of the constant z.

Solution

$P = 690$ MPa, $H/D = 0.8$, $\mu = 0.3$, $\sigma = ?$

By using Equation 12.14,

$$\sigma = P[1 - \mu z(H/D)] = 690 [1 - (0.3 \times z \times 0.8)] = 690(1 - 0.24z)$$

The average compaction stress, MPa $= 690(1 - 0.24z)$.

EXAMPLE 12.12: CALCULATING THE FILLING DEPTH IN METAL-POWDER COMPACTION

An iron powder has an apparent density of $2.5\,g/cm^3$, and a compact density of $6.4\,g/cm^3$. The height of the compact is 20 mm. Calculate the filling depth.

Solution

$\rho_c = 6.4\,g/cm^3$; $\rho_a = 2.5\,g/cm^3$; $H = 20\,mm = 2\,cm$; Filling depth $= d_f = ?$

By using Equation 12.15,

$$\rho_c/\rho_a = d_f/H$$

$$6.4/2.5 = d_f/2$$

$d_f = 5.12\,cm = 51.2\,mm.$

EXAMPLE 12.13: CALCULATING PERCENT SHRINKAGE IN SINTERED COMPACT OF METAL POWDER

A P/M compact was sintered, and the length before and after sintering was measured to be 4 cm and 3.72 mm, respectively. Calculate the percent shrinkage.

Solution

$l_s = 4\,cm$, $\Delta l = 4.00 - 3.72 = 0.28\,cm.$

By using Equation 12.16,

$$\% \text{ Shrinkage} = [(\Delta l)/l_s] \times 100 = (0.28/4) \times 100 = 7$$

Shrinkage $= 7\%.$

Questions and Problems

12.1 Briefly explain the main steps in P/M with the aid of a sketch.

12.2 List the various techniques of metal powder production, and explain one of them with the aid of a sketch.

12.3 (a) List the various techniques of metal powder characteristics. (b) Sketch a mesh screen showing it is helpful in measuring particle size.

12.4 (a) Draw a sketch showing various particle shapes of metal powders. (b) Compare angular-shaped particles with spherical-shaped particles with reference to their advantages and disadvantages.

12.5 How are flow rate and apparent density of metal powders determined?

12.6   Briefly explain the metal-powder compaction with the aid of a sketch.

P12.7   Why is the strength of sintered compact superior as compared to "green compact"?

P12.8   A metal powder was produced by using inert-gas atomization technique. The metal-droplets temperature was 1900°C; the diameter of a droplet was measured to be 15 μm. The inert gas was injected into the atomization chamber at a temperature of 600°C. The convective heat transfer coefficient for the metal powder production system is $0.7\,J/s\,cm^2$. Calculate the *cooling rate* of the metal droplets. The metal powder data: density $= 8.2\,g/cm^3$; $c_{pm} = 0.48\,J/g\,K$.

P12.9   By using the data in P12.8, calculate the *SDAS* for the powder particles.

P12.10   The rod length and width of an acicular metal-powder particle are 1.5 mm and 0.4 mm, respectively. Calculate its: (a) aspect ratio and (b) surface area.

P12.11   The particle size distribution data for a metal powder are given in Table below:

Particle size, μm	1.9	2.4	4.8	10.4	20.0	45.6	57.3	70.8	80.7
Cumulative % undersize	0.3	0.8	2.7	9.1	46.8	78.6	84.4	92.8	100

   Estimate the (a) arithmetic mean, (b) quadratic mean, and (c) cubic mean.

P12.12   There are 32,400 openings per square inch in a mesh screen. Calculate the following: (a) the mesh count and (b) the length of a mesh in the screen.

P12.13   A 95 wt% iron powder was blended with 3 wt% graphite and 2 wt% impurity (with density $8\,g/cm^3$). Calculate the theoretically achievable pore-free density of the powder mix.

P12.14   A metal powder was tested in a *Hall flowmeter* and then weighed; the mass of the powder was measured to be 13.51 g. Calculate the apparent density of the powder.

P12.15   A metal powder has an apparent density of $2.7\,g/cm^3$, and a compact density of $6.8\,g/cm^3$. The height of the compact is 22 mm. Calculate the filling depth.

P12.16   A *P/M* compact was sintered and the length before and after sintering was measured to be 5.00 cm and 4.80 mm, respectively. Calculate the percent shrinkage.

References

Anwar, M.A. 2017. Powder Metallurgy. In: Z. Huda (Ed.); *Materials Processing for Engineering Manufacture*. Zurich, Switzerland: Trans Tech Publications, 223–251.

Askeland, D.R., Wright, W.J. 2016. *The Science and Engineering of Materials*. Independence, KY: Cengage.

Hofmann, H., Bowen, P. 2017. Internet Source: http://ltp.epfl.ch/files/content/sites/ltp/files/shared/Teaching/Master/06PowderTechnology/Compaction.pdf; Accessed on October 4.

Höganäs, A.B. (Ed.) 2013. Production of Sintered Components. In: *Höganäs Handbook for Sintered Components*. Sweden: Höganäs, AB, 7–158.

Samal, P.K., Newkirk, J.W. (Eds.). 2015. *ASM Handbook, Vol. 7: Powder Metallurgy*. Materials Park, OH: ASM International.

Schulz, G. 2001. Ultrafine metal powders for high temperature applications made by gas atomization. In: Kneringer, G., Roedhammer, P., Wildner, H. (Eds.); *Proceedings 15th International Plansee Seminar*; Reutte (Austria); 74–84.

Shiwen, H., Yong, L., Sheng, G. 2009. Cooling rate calculation of non-equilibrium aluminum alloy powders prepared by gas atomization. *Rare Metal Materials and Engineering*, 38(1), 353–356.

Upadhyaya, A., Upadhyaya, G.S. 2011. *Powder Metallurgy: Science, Technology, and Materials*. Hyderabad, India: Universities Press.

13

Weld Design and Joining

13.1 Welding, Brazing, and Soldering

13.1.1 Distinction between Welding, Brazing, and Soldering

Welding involves the joining of two or more parts by *coalescence* of the contacting surfaces by a suitable application of heat and/or pressure. What is special in *welding* (fusion welding) is that it involves *coalescence* of the contacting surfaces by melting of base metals—a feature that is absent in brazing and soldering. Welded joints are stronger than brazed and soldered joints (Althouse and Turnquist, 2012). Figure 13.1 illustrates an electric arc welding process (a form of *fusion welding*); here the base metal is heated above its melting temperature to form a weld joint.

In brazing and soldering, a filler is melted to form the joint (see Figure 13.2). *Brazing* is a joining process in which a filler metal is melted and distributed by capillary action between the faying surfaces of the metal parts. In brazing, the filler metal has a melting temperature above 450°C but below the melting temperature of the base metal (Zaharinie et al., 2015). *Soldering* is a joining process in which a filler metal with melting point below 450°C is melted and distributed between the faying surfaces of the metal parts.

13.1.2 Mathematical Modeling of Brazing and Soldering

Braze Joint Design: The design of a braze joint requires considerations of shear strength of brazing filler metal as well as the tensile strength of the weakest member (base metal); these considerations can be mathematically expressed as follows (Lucas-Millhaupt, 2011–2017):

$$l = (S_{ut} \cdot s)/(C \cdot \tau) \tag{13.1}$$

where l is the length of lap, mm; S_{ut} is the ultimate tensile strength of the weakest member, MPa; s is the thickness of the weakest member, mm; C is the joint integrity factor of 0.8; and τ is the shear strength of brazed filler metal (see Figure 13.3).

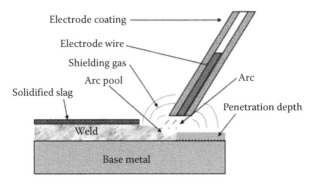

FIGURE 13.1
Arc welding (shielded metal arc welding) process.

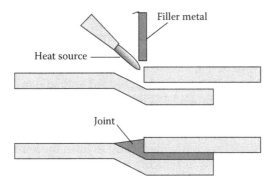

FIGURE 13.2
Brazing/soldering principle.

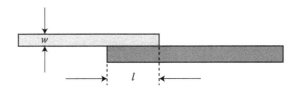

FIGURE 13.3
The length of lap (*l*) and the thickness (*W*) for brazing.

Role of Wetting in Brazing/Soldering: The success of a soldering process is mainly achieved by an *effective spreading* or *wetting* of liquid filler metal between the faying surfaces. The effective *wetting* strongly depends on the contact angle, θ; the other soldering parameters include the surface tension between solid metal and air (γ_s), the surface tension between liquid solder and air (γ_L), and the surface tension between solid metal and liquid solder (γ_{SL}) (see Figure 13.4).

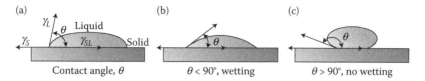

FIGURE 13.4
The dependence of *wetting* on the contact angle, θ. (a) Actions of the surface tension between solid metal and liquid solder, γ_{SL}, the surface tension between liquid solder and air, γ_L, and the surface tension between solid metal and air, γ_S, (b) acute contact angle results in effective wetting, and (c) obtuse contact angle results in no wetting.

The surface tension between solid metal and liquid solder (γ_{SL}) is related to the contact angle by

$$\gamma_{SL} = \gamma_S - \gamma_L \cos\theta \tag{13.2}$$

It is evident in Figure 13.4 that in order to achieve an effective spreading or *wetting*, the γ_{SL} value should be minimized. This objective of low γ_{SL} value is achieved by ensuring that the contact angle is acute (i.e., $\theta < 90°$), since the cosine of an obtuse angle is a negative value, which means a high γ_{SL} (see Equation 13.2).

13.2 Weld Joints Design

Principles: The selection of a correct weld joint design is critical to the successful fabrication of metals and alloys. In particular, a welded pressure vessel/boiler may be a cause of fatal injuries and loss of properties if they are based on a poor weld design. A poor joint design may result in negation of even the most optimum welding conditions. An important consideration in weld joint design of high-strength alloys is to provide sufficient accessibility and space for movement of the welding electrode or filler metal. In particular, a larger included weld angle, wider root opening (gap), and reduced land (root face) thickness are typically required in superalloys welding as compared to welding with steels. Figure 13.5 illustrates the five types of weld joints commonly practiced in the fabrication industry.

Mathematical Modeling: High-strength applications require butt weld joints. Table 13.1 presents various butt weld orientations and the corresponding throat area formulas.

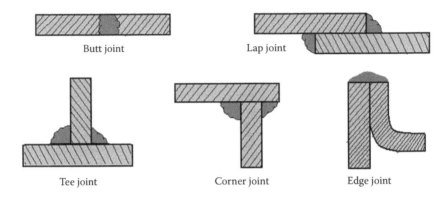

FIGURE 13.5
The five types of weld joint designs.

TABLE 13.1

Throat Areas for Various Butt Weld Orientations

Butt Weld Orientation	Throat Area of Butt Weld along Full Length	
Butt end-tube weld	$A = s \cdot l$	(13.3)
Circumferential tube butt weld	$A = \pi s(D - t)$	(13.4)
Butt weld angled relative to applied load	$A = s(l/\cos \delta)$	(13.5)
Butt weld normal to applied load	$A = s \cdot l$	(13.6)

The meanings of the terms in Table 13.1 are as follows: A is the throat area mm^2; s is the thickness of a thinner joined part, mm; l is the weld length, mm; D is the tube outer diameter, mm; and δ is the weld angle (deg).

Stresses in butt welds are determined on the basis of whether they are in tension or shear (Figure 13.6).

The *tensile stress* in butt weld can be calculated by using an American Institute of Steel Construction (AISC) code as follows:

$$\sigma = (F/hl) \leq 0.6S_y \tag{13.7}$$

where σ is the tensile stress, MPa; F is the tensile force, N; h is the weld leg, mm; l is the weld length, mm; and S_y is the yield strength of the base metal, MPa.

The shear stress in butt weld can be calculated by using the AISC code as follows:

$$\tau = (F/hl) \leq 0.3S_{ut} \tag{13.8}$$

where τ is the shear stress, MPa; and S_{ut} is the ultimate tensile strength of the base metal, MPa.

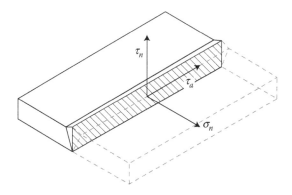

FIGURE 13.6
Stresses in a butt weld (σ_n = tensile stress normal to the weld axis, τ_a = shear stress along the weld axis, and τ_n = shear stress normal to the weld axis).

13.3 Design of Fillet Welds

A *fillet weld* is a type of weld that has a roughly triangular cross section (see Figure 13.6). The design of a fabricated structure must be based on avoidance of stress concentration sites, which may be ensured by providing fillet welds (see Figure 13.7). A fillet weld always fails by *shear*.

The shear stress in the fillet weld (Figure 13.7) can be computed by using the AISC code as follows:

$$\tau = F/A = F/(0.707hl) \leq 0.3S_{ut} \tag{13.9}$$

where τ is the shear stress, MPa; F is the force acting parallel to the weld length, N; and A is the shear area of the fillet welds, mm^2; h is the weld leg, mm; and l is the weld length, mm. For double parallel fillet welds, the total shear area of the fillet welds, $A = 2 \times 0.707hl = 1.414hl$.

The significance of Equation 13.9 is illustrated in Examples 13.8 through 13.10.

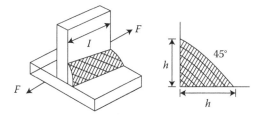

FIGURE 13.7
Parallel fillet welds.

13.4 Welding Processes

13.4.1 Principles and Classification of Welding

All welding processes may be broadly classified into two groups: fusion welding and solid-state welding; each group has subgroups of welding processes. This section introduces important welding processes in both fusion and solid-state welding. In *fusion welding*, the base metals are heated above their melting temperatures to cause *coalescence* resulting in a *fusion weld*. In *solid-state* welding, there is a combination of heat and pressure, but the temperature is less than the melting temperature of the base metal. *Fusion welding* processes include oxyfuel welding (OFW), Thermit welding (TW), shielded metal arc welding (SMAW), submerged arc welding (SAW), gas tungsten arc welding (GTAW), gas metal arc welding (GMAW), flux core arc welding (FCAW), *RSW*, resistance seam welding (RSeW), electron beam welding, and laser beam welding. Important processes in solid-state welding include cold pressure welding (CPW), hot pressure welding, friction welding (FRW), explosive welding (EXW), ultrasonic welding (USW), and the like (Cary and Helzer, 2004). The commonly practiced welding processes are briefly explained in the following sections.

13.4.2 Fusion Welding Processes

Fusion welding was introduced the preceding section. For a good quality fusion weld, it is desirable to melt the metal with minimum energy and high *power density*. *Power density* (*PD*) is defined as the power transferred to the work per unit surface area. Mathematically, *PD* is expressed by (Groover, 2010)

$$PD = P/A \tag{13.10}$$

where *PD* is the power density, W/mm^2; *P* is the power entering the surface, *W*; and *A* is the surface area over which the energy is entering, mm^2 (see Example 13.14).

13.4.2.1 Arc Welding

13.4.2.1.1 Principles and Mathematical Models in AW

Principles: AW involves heating and melting of the material by an electric arc that generates a very high temperature of about 5000°C that results in a molten pool and forms a weld joint. Figure 13.8 illustrates the basic equipment setup for AW; here an electrode cable supplies current to the electrode, whereas an earth cable connects the workpiece to the welding machine to provide a return path for the current.

   Mathematical Models: The quality of an arc weld strongly depends on the heat input during welding. *Heat input* is the energy transferred per unit length of weld. In general, a lower heat input results in a better weld quality.

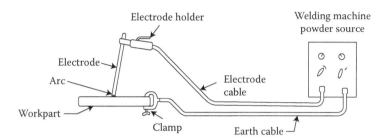

FIGURE 13.8
Equipment setup for electric AW.

Funderburk has reported useful mathematical models based on the heat input required in AW; these models are presented in the following paragraphs (Funderburk, 1999).

The *heat input* required in electric arc welding can be computed by

$$H_i = (0.06 \cdot I \cdot V)/v_w \qquad (13.11)$$

where H_i is the *heat input*, kJ/mm; I is the current, A; V is the arc voltage, volts; and v_w is the welding speed, mm/min.

The fillet weld leg size is approximately related to heat input by

$$h = (H_i/500)^{1/2} \qquad (13.12)$$

where h is the fillet-weld leg size, mm; and H_i is the *heat input*, kJ/mm.

The cooling rate of an arc weld is a primary factor that determines the final metallurgical structure of the weld and heat-affected zone (HAZ). For a good quality weld, a faster cooling rate is desirable. The cooling rate (dT/dt) can be computed by

$$dT/dt = C/(T_o H) \qquad (13.13)$$

where C is the constant of proportionality; T_o is the pre-heat temperature; and H is the heat input.

13.4.2.1.2 Arc Welding Processes

Shielded Metal Arc Welding: The SMAW process involves the use of a stick electrode that is covered with an extruded coating of flux. The heat of the arc melts the flux coating that generates a gaseous shield to protect the molten pool against oxidation (see Figure 13.1). The flux ingredients react with unwanted impurities producing a slag that floats to the surface of the weld pool and forms a crust that protects the weld during cooling. When the weld is cold, the slag is chipped off.

Submerged Arc Welding: The SAW process involves formation of an invisible arc between a continuously fed bare wire electrode and the workpiece by using a flux to generate protective gases and slag, and to add alloying elements to the weld pool.

Gas Tungsten Arc Welding: GTAW is an AW process that uses a nonconsumable tungsten electrode to weld a metal that is protected by an inert gas. In general, a filler wire is added to the weld pool separately.

Gas Metal Arc Welding: The GMAW process involves the use of a metal wire (consumable electrode) to strike an arc between the wire and the workpiece. The electrode wire is fed continuously through a welding torch into the weld pool; an inert gas is also admitted along the tube and into the torch and exits around the wire.

Flux Core Arc Welding: The FCAW process involves the uses of a tubular wire that is filled with a flux. Direct current, electrode positive (DCEP) is commonly employed, which results in the striking of an arc between the continuous wire electrode and the workpiece. The flux melts during welding and shields the weld pool from the atmosphere.

13.4.2.2 Other Fusion Welding Processes

Oxyfuel Welding: OFW is the joining of metals by heating them with a fuel gas (usually acetylene) flame with the use of filler metal. This process involves the melting of the base metal and a filler metal by means of the flame produced at the tip of a welding torch.

Thermit Welding: TW is a fusion welding process that uses thermite material composition and exothermic reaction to heat and melt the metal. The exothermic reaction occurs between aluminum powder and a metal oxide to form alumina and molten iron to form a weld.

Resistance Spot Welding: The RSW process uses pointed copper electrodes to provide the passage of current as well as the transmission of pressure that is required for the formation of strong *spot welds*. In RSW, two or more overlapped metal sheets are joined by spot welds due to the combined effects of heat and pressure. The diameter of the weld spot (weld nugget) is in the range of 3–12 mm (see Figure 13.9).

The heat generated in RSW can be calculated by the following formula:

$$Q_{RSW} = I^2 R t \tag{13.14}$$

where Q_{RSW} is the heat energy generated during RSW, J; I is the current, A; R is the effective resistance during the spot welding operation, Ω; and t is the time, s (see Example 13.16).

The heat required to melt the weld nugget can be determined by

$$Q_m = m H_f \tag{13.15}$$

where Q_m is the heat energy required to melt the weld nugget, J; m is the mass of the weld nugget, g; and H_f is the heat of fusion of the work metal, J/g (see Example 13.18).

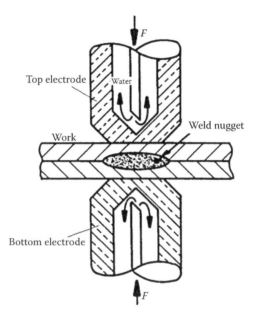

FIGURE 13.9
Schematic of *RSW* process.

The mass of the nugget (*m*) can be calculated when the density of the work metal and volume of the nugget are known. The volume of the nugget can be computed by (see Figure 13.9)

$$V_{nugget} = \text{Projected volume between the electrodes} = (\pi/4)D^2\,x \quad (13.16)$$

where V_{nugget} is the volume of the nugget, mm³; D is the diameter of the electrode tip, mm; and x is the nugget thickness, mm. The nugget thickness (x) can be estimated by

$$x \cong 0.8X = 0.8(2s) = 1.6s \quad (13.17)$$

where X is the distance between the electrodes, mm; and s is work (sheet) thickness, mm (see Example 13.17). Once the values of Q_{RSW} and Q_m are known, the heat dissipated to the metal surrounding the nugget can be calculated by

$$Q_{lost} = Q_{RSW} - Q_m \quad (13.18)$$

where Q_{lost} is the heat energy lost to the surroundings, J (see Example 13.19).

Resistance Seam Welding: The RSeW process involves the use of disc (wheel)-shaped electrodes that rotate as the material passes between them. The applications of direct current (*d.c.*) and the pressure provided by the wheels are sufficient to generate heat and pressure resulting in the production of a series of leak-tight welds.

13.4.3 Solid-State Welding Processes

Cold Pressure Welding: CPW involves the application of high pressures (1400–2800 MPa for aluminum and at least double the value for copper) at ambient temperature. The high pressure results in interfacial deformations of around 70% that break the surface oxide layer to expose fresh, uncontaminated metal that makes a weld joint.

Explosive Welding: The EXW involves the use of an explosive to detonate the surface of a metal resulting in the generation of a high-pressure pulse, which propels the clad metal at a very high speed. The fast-moving clad metal hits the base metal at an angle causing welding.

Ultrasonic Welding: USW involves the use of a welding horn to generate mechanical vibrations above the audible range for welding. The vibrations are used to soften or melt the thermoplastic (base) material at the joint line.

Friction Welding: FRW involves the production of a weld under friction (rapid rotation of one piece against a stationary piece) and compression force. FRW is a suitable choice of welding metallic workpieces with a rotational symmetry (e.g., round rods, bars, tubes, etc.).

13.5 Problems and Solutions—Examples in Weld Design and Joining

EXAMPLE 13.1: DETERMINING THE EXTENT OF WETTING IN SOLDERING

Determine the extent of wetting in a soldering operation for the following values of contact angles: (a) $\theta = 100°$, (b) $\theta = 45°$, and (c) $\theta = 0$.

Solution

(a) By using Equation 13.2,

$$\gamma_{SL} = \gamma_S - \gamma_L \cos \theta = \gamma_{SL} = \gamma_S - \gamma_L \cos 100°$$
$$= \gamma_S - (-0.17)\, \gamma_L = \gamma_S + 0.17\gamma_L$$

This calculation results in a high value of γ_{SL}, which means no wetting.

(b) $\gamma_{SL} = \gamma_S - \gamma_L \cos \theta = \gamma_{SL} = \gamma_S - \gamma_L \cos 45° = \gamma_S - 0.707\gamma_L$

This calculation results in an intermediate value of γ_{SL}, which means a good wetting.

(c) $\gamma_{SL} = \gamma_S - \gamma_L \cos \theta = \gamma_S - \gamma_L \cos 0 = \gamma_S - \gamma_L$

> This calculation results in the lowest value of γ_{SL}, which means the best wetting.

EXAMPLE 13.2: DETERMINATION OF THROAT AREA IN CIRCUMFERENTIAL TUBE BUTT WELD

Two circumferential tubes each of outer diameter 2 cm are circumferentially butt welded. The thickness of a thinner joined tube is 3 mm. Calculate the throat area of the full-length butt weld.

Solution

$D = 2$ cm $= 20$ mm, $s = 3$ mm, $A = ?$

By using Equation 13.4,

$$A = \pi t(D - s) = \pi \times 3(20 - 3) = 160.2 \text{ mm}^2$$

The throat area of the full-length butt weld $= 160.2 \text{ mm}^2$.

EXAMPLE 13.3: DETERMINATION OF THROAT AREA IN BUTT WELD NORMAL TO APPLIED LOAD

Two metal plates each of thickness 4 mm are butt welded to form a 5 cm long welded joint such that the butt weld is normal to applied load. Calculate the throat area.

Solution

$s = 4$ mm, $l = 5$ cm $= 50$ mm, $A = ?$

By using Equation 13.6,

$$A = s \cdot l = 4 \times 50 = 200 \text{ mm}^2$$

The throat area $= 200 \text{ mm}^2$.

EXAMPLE 13.4: DETERMINATION OF THROAT AREA IN BUTT WELD ANGLED RELATIVE TO APPLIED LOAD

Two metal plates each of thickness 5 mm are butt welded to form a 7 cm long welded joint such that the butt weld is at an angle of 60° to the applied load. Calculate the throat area.

Solution

$s = 5$ mm, $l = 7$ cm $= 70$ mm, $\delta = 60°$, $A = ?$

By using Equation 13.5,

$$A = s(l/\cos \delta) = 5(70/\cos 60°) = 5(70/0.5) = 700 \text{ mm}^2$$

The throat area $= 700 \text{ mm}^2$.

EXAMPLE 13.5: CALCULATING THE TENSILE STRESS IN A BUTT WELDED PART

A tensile force of 60 kN acts on butt-welded base plates according to Figure 13.4(a). The weld leg is 5 mm and the weld length is 55 mm. Calculate the tensile stress.

Solution

$F = 60,000\,$N, $h = 5\,$mm, $l = 55\,$mm.

By using Equation 13.7,

$$\sigma = (F/hl) = (60,000)/(5 \times 55) = 60,000/275 = 218\,\text{N/mm}^2$$

The tensile stress $= 218\,$MPa.

EXAMPLE 13.6: DETERMINING THE SAFETY IN THE DESIGN OF THE WELDED PART UNDER TENSILE STRESS

By using the data in Example 13.5, determine the safety of the design of the welded plates if the yield strength of the plate material is 400 MPa.

Solution

According to Equation 13.7,

$$\sigma \leq 0.6 \quad S_y = 0.6 \times 400 = 240\,\text{MPa}$$

Since the tensile stress, σ, as determined in Example 13.5 is 218 MPa, which is <240, the design of the welded plates is safe.

EXAMPLE 13.7: DETERMINING THE SAFETY IN THE DESIGN OF THE WELDED PART UNDER SHEAR STRESS

A shear stress of 500 MPa acts on butt-welded plates. Will failure occur if the ultimate strength of the plate material is 800 MPa?

Solution

$\tau = 500\,$MPa, $S_{ut} = 800\,$MPa.

According to Equation 13.8, the butt-welded design is safe if $\tau < 0.3\,S_{ut}$.

$$0.3 S_{ut} = 0.3 \times 800 = 240\,\text{MPa}$$

Since 500 MPa > 240 MPa, the design of the welded plates is *not* safe.

EXAMPLE 13.8: CALCULATING THE SHEAR STRESS IN A FILLET WELD

A parallel fillet welded unit (see Figure 13.7) is subjected to force, $F = 65$ kN. The weld leg is 6 mm and the weld length is 58 mm. Calculate the shear stress acting on the fillet weld.

Solution

$F = 65{,}000$ N, $h = 6$ mm, $l = 58$ mm, $\tau = ?$

By using Equation 13.9,

$$\tau = F/(0.707hl) = 65{,}000/(0.707 \times 6 \times 58) = 65{,}000/246 = 264.2 \text{ MPa}$$

The shear stress acting on the fillet weld $= 264.2$ MPa.

EXAMPLE 13.9: DESIGNING A FILLET WELD AND AW IN ACCORDANCE WITH AISC WELDING CODE

The end P of the (6 mm $\times$ 80 mm) flat plate is double parallel fillet welded to a vertical post as shown in Figure E13.9.
(a) What are the limitations on the geometric sizes of fillet welds at P?
(b) Design the fillet welds at P according to the AISC welding code (Table E13.9) and select the most economical electrode (i.e., the electrode of least strength).

Solution

(a) Geometric sizes of the fillet welds at P: $h \leq 6$ mm, $l = 70$ mm.
(b) Force, $F = 78$ kN $= 78{,}000$ N, Total shear area of double parallel fillet welds, $A = 1.41hl$.

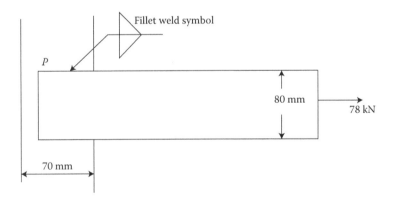

FIGURE E13.9
Vertical post double fillet welded to the plate.

TABLE E13.9

Welding Electrode Strengths Using AISC Welding Code

	E60xx	E70xx	E80xx	E100xx	E110xx
Yield strength, S_y, MPa	345	390	470	600	670
Ultimate tensile strength, S_{ut}, MPa	427	482	550	698	760

By using Equation 13.9, the applied shear stress can be computed as follows:

$$\tau = F/A = (78{,}000)/(1.41hl) = 55{,}319/hl \qquad \text{(E13.9a)}$$

Again, by using Equation 13.9,

$$\tau = 0.3S_{ut} \qquad \text{(E13.9b)}$$

By combining Equations E13.9a and E13.9b

$$55{,}319/hl = 0.3S_{ut}$$

$$hl = 184{,}397/S_{ut}$$

By reference to Table E13.9, we first take E60xx as the most economical electrode since it has the lowest $S_{ut} = 427$ MPa.

$$hl = 184{,}397/427 = 432 \text{ mm}^2$$

However, the limitations on the fillet geometry say $hl \leq 6 \times 70 \leq 420$ mm^2.

Since 432 mm$^2 > 420$ mm^2, the use of E60xx electrode is not recommended.

Now, we try the next economical electrode—that is, E70xx for which $S_{ut} = 482$ MPa:

$$hl = 184{,}397/\sigma_U = 184{,}397/482 = 382.5 \text{ mm}^2$$

Since 382.5 mm$^2 < 420$ mm^2, E70xx is the most economical electrode.

Hence, the fillet weld design is as follows: $h = 6$ mm, $l = 70$ mm, E70xx electrode is selected.

EXAMPLE 13.10: CALCULATING THE HEAT INPUT IN ELECTRIC ARC WELDING

A steel part was arc welded at a welding speed of 20 cm/min by passing a current of 175 A at arc voltage of 90 volts. Calculate the required heat input for *AW*.

Solution

$I = 175$ A, $V = 90$ volts, $v_w = 20$ cm/min $= 200$ mm/min, $H_i = ?$

By using Equation 13.11,

$$H_i = (0.06 \cdot I \cdot V)/v_w = (0.06 \times 175 \times 90)/200 = 4.7 \text{ kJ/mm}$$

The heat input $= 4.7$ kJ/mm.

EXAMPLE 13.11: CALCULATING THE FILLET WELD LEG SIZE WHEN THE HEAT INPUT IS KNOWN

By using the data in Example 13.10, calculate the fillet weld leg size for the AW process.

Solution

$H_i = 4.7$ kJ/mm.

By using Equation 13.12,

$$h = (H_i/500)^{1/2} = (4.7/500)^{1/2} = 0.09 \cong 0.1 \text{ mm}$$

The fillet weld leg $= 0.1$ mm.

EXAMPLE 13.12: DETERMINATION OF FILLET WELD LEG SIZE WHEN THE WELDING SPEED IS HALVED

By using the data in Example 13.11, compute the fillet weld leg size if welding speed is reduced by one-half during AW. Will the new fillet weld size be a better design?

Solution

$I = 175$ A, $V = 90$ volts, $v_w = 10$ cm/min $= 100$ mm/min.

By using Equation 13.11,

$$H_i = (0.06 \cdot I \cdot V)/v_w = (0.06 \times 175 \times 90)/100 = 9.45 \text{ kJ/mm}$$

By using Equation 13.12,

$$h = (H_i/500)^{1/2} = (9.45/500)^{1/2} = 0.14 \text{ mm}$$

There is not a big difference between the previous and new fillet weld leg sizes. The new (reduced) welding speed is extremely undesirable because it results in the heat input that is too high. Hence, the welding process parameters in Example 13.11 are better.

EXAMPLE 13.13: DETERMINING THE CONSTANT OF PROPORTIONALITY FOR COOLING RATE OF ARC WELD

A heat input of 2.5 kJ/mm was used in AW of a steel section that was preheated at a temperature of 100°C. The cooling rate of the arc weld was measured to be 7.8°C/s. Calculate the value of the constant of proportionality in the cooling rate expression.

Solution

By using Equation 13.13,

$$dT/dt = C/(T_oH)$$

$$7.8 = C/(100 \times 2.5)$$

$$C = 250 \times 7.8 = 1950 \ (kJ \cdot {}^\circ C)/(mm \cdot s)$$

The constant of proportionality for cooling rate of the arc weld $= 1950$ $(kJ \cdot {}^\circ C)/(mm \cdot s)$.

EXAMPLE 13.14: CALCULATING THE POWER DENSITIES IN VARIOUS REGIONS DURING FUSION WELDING

A fusion welding heat source transfers 2800 W to the surface of a metal part, which is subjected to varying heat intensities as one moves from the surface to inside the metal. Then 75% of the power is transferred within a circle of diameter 4 mm, and 92% of the power is transferred within a concentric circle of diameter 10 mm. Calculate the *power densities* in (a) the 4 mm diameter inner circle and (b) the 10 mm diameter ring that lies around the inner circle.

Solution

Power $= P = 2800$ W; Diameter of the inner circle $= d_i = 4$ mm, Diameter of ring $= d_r = 10$ mm.

(a) The surface area of the inner circle $= A_i = (\pi d_i^2)/4 = (\pi \times 4^2)/4 = 12.6$ mm^2.

   The power in the inner circle area $= P_i = 75\% \times P = 0.75 \times 2800 = 2100$ W.

   By using Equation 13.10,
   Power density for the inner circle $= (PD)_i = P_i/A_i = 2100/12.6 = 166.67$ W/mm^2

   The power density for the 4 mm diameter inner circle $= 166.67$ W/mm^2.

(b) The surface area of the ring outside the inner circle $= A_r = [\pi (d_r^2 - d_i^2)]/4 = [\pi (10^2 - 4^2)]/4$ $A_r = [\pi (100 - 16)]/4 = 66$ mm^2.

   The power in the ring $= P_r = (92\% \ P) - 2100 = (0.92 \times 2800) - 2100 = 476$ W.

   Power density for the inner circle $= (PD)_r = P_r/A_r = 476/66 = 7.2$ W/mm^2.

EXAMPLE 13.15: CALCULATING THE LENGTH OF LAP FOR FLAT BRAZE JOINTS

Calculate the length of lap that is needed to join a 1.3 mm thick annealed Monel metal (tensile strength $= 482.6$ MPa) sheet to a metal of greater strength. The shear strengths for silver brazing filler metals are typically around 172.4 MPa.

Solution

Tensile strength of the weakest metal (Monel) = S_{ut} = 482.6 MPa, s = 1.3 mm, τ = 172.4 MPa.

By using Equation 13.1,

$$l = (S_{ut} \cdot s)/(C \cdot \tau) = (482.6 \times 1.3)/(0.8 \times 172.4) = 4.5 \text{ mm}$$

The length of lap for brazing = 4.5 mm.

EXAMPLE 13.16: CALCULATING THE HEAT GENERATED IN *RSW*

Two 1.2 mm thick steel sheets are spot welded by using a current of 6000 A. The current flow time duration is 0.2 s, and each electrode tip has a diameter of 6 mm. The effective resistance in the *RSW* operation is 150 µ Ω. Calculate the heat generated in producing the spot weld.

Solution

I = 6000 A, t = 0.2 s, R = 150 µ Ω = 150×10^{-6} Ω = 0.00015 Ω.

By using Equation 13.14,

$$Q_{RSW} = I^2 R t = (6000)^2 \times 0.00015 \times 0.2 = 1080 \text{ J} = 1.08 \text{ kJ}$$

The heat energy generated in producing the spot weld = 1.08 kJ.

EXAMPLE 13.17: CALCULATING THE VOLUME OF WELD NUGGET IN *RSW*

By using the data in Example 13.16, calculate the volume of the weld nugget in *RSW*.

Solution

Work (sheet) thickness = s = 1.2 mm, D = 6 mm.

By using Equation 13.17,

$$x \cong 0.8X = 0.8(2s) = 1.6s = 1.6 \times 1.2 = 1.92 \text{ mm}$$

By using Equation 13.16,

$$V_{nugget} = (\pi/4)D^2 x = 0.785 \times 6^2 \times 1.92 = 54.3 \text{ mm}^3$$

The volume of the weld nugget = 54.3 mm^3.

EXAMPLE 13.18: CALCULATING THE HEAT REQUIRED TO MELT THE WELD NUGGET

By using the data in Example 13.17, calculate the heat required to melt the weld nugget. The latent heat of fusion of steel is 1400 J/g.

Solution

$V_{nugget} = 54.3$ mm^3; Density of steel $= \rho_{steel} = 8$ g/cm$^3 = 0.008$ g/mm^3, $H_f = 1400$ J/g, $Q_m = ?$

$$\rho_{steel} = m/V_{nugget}$$

$$m = \rho_{steel} \times V_{nugget} = 0.008 \times 54.3 = 0.43 \text{ g}$$

By using Equation 13.15,

$$Q_m = mH_f = 0.43 \times 1400 = 608 \text{ J}$$

The heat required to melt the weld nugget $= 608$ J.

EXAMPLE 13.19: COMPUTING THE HEAT DISSIPATED INTO THE METAL SURROUNDING THE WELD NUGGET

By using the data in Examples 13.16 and 13.18, calculate the heat dissipated into the metal surrounding the weld nugget. What percentage of the generated heat energy is lost to the surrounding?

Solution

$Q_m = 608$ J, $Q_{RSW} = 1080$ J, $Q_{lost} = ?$

By using Equation 13.18,

$$Q_{lost} = Q_{RSW} - Q_m = 1080 - 608 = 472 \text{ J}$$

Percentage energy lost $= [(Q_{RSW} - Q_m)/Q_{RSW}] \times 100 = (472/1080) \times 100 = 43.7\%$.

The heat dissipated to the metal surrounding the weld nugget $= Q_{lost} = 472$ J.

Energy lost $= 43.7\%$.

Questions and Problems

13.1 Make a distinction between welding, brazing, and soldering with the aid of diagrams.

13.2 What is the difference between fusion welding and solid-state welding?

13.3 Classify the following welding processes under the groups: (a) fusion welding and (b) solid state welding: OFW, SMAW, FRW, RSW, USW, GTAW, TW, and CPW.

13.4 Draw a diagram showing the various weld joints. Which is the strongest weld joint?

13.5   Explain electric arc welding with the aid of a diagram.

P13.6   A heat input of 2.2 kJ/mm was used in AW of a steel section that was preheated at a temperature of 107°C. The cooling rate of the arc weld was measured to be 8°C/s. Calculate the value of the constant of proportionality in the cooling rate expression.

P13.7   Determine the extent of wetting in a soldering operation for the following values of contact angles: (a) $\theta = 10°$ and (b) $\theta = 120°$.

P13.8   Two metal plates each of thickness 4 mm are butt welded to form a 6 cm long welded joint; the butt weld is at an angle of 50° to the applied load. Calculate the throat area.

P13.9   A parallel fillet welded unit (see Figure 13.5) is subjected to force, $F = 70$ kN. The weld leg is 7 mm and the weld length is 62 mm. Calculate shear stress acting on the fillet weld.

P13.10   By using the data in Example 13.10, design the fillet weld and select the most economical electrode if the force = 75 kN.

P13.11   A steel part was arc welded at a welding speed of 25 cm/min by passing a current of 180 A at arc voltage of 95 volts. Calculate the required heat input for AW.

P13.12   By using the data in P13.11, calculate the fillet weld leg size for the AW process.

References

Althouse, A.D., Turnquist, C.H. 2012. *Modern Welding*, 11th Edition. Tinley Park, IL: Goodheart-Willcox.

Cary, H.B., Helzer, S. 2004. *Modern Welding Technology*, 6th Edition. New York: Pearson.

Funderburk, S.R. 1999. Key concepts in welding engineering. *Welding Innovation*, XVI (1), 1–4.

Groover, M. 2010. *Fundamentals of Modern Manufacturing*, 4th Edition. New York: Wiley.

Lucas-Millhaupt. 2011–2017. The Principles of Joint Design. Internet Source: https://www.brazingbook.com/the-principles-of-joint-design/

Zaharinie, T., Huda, Z., Izuan, M.F., Hamdi, M. 2015. Development of optimum process parameters and a study of the effects of surface roughness on brazing of copper. *Applied Surface Science*, 331(3), 127–131.

14

Glass and Ceramic Processing

14.1 Glass Temperature-Sensitive Properties and Mathematical Modeling

Glass transition temperature: An important thermal property of glass is its glass transition temperature, T_g; it is defined as the temperature range over which the reversible transition of glass from a hard and relatively brittle "glassy" state into a viscous or rubbery state occurs as the temperature is increased.

Viscosity-temperature curves: It is considered that a liquid on being cooled becomes practically a glass when its viscosity becomes equal to 10^{12} poises (Ojovan, 2008). Glass-forming operations involve a great deal of viscosity-temperature characteristics of the glass being processed; this material behavior is illustrated in viscosity versus temperature curves (Figure 14.1).

It is evident in Figure 14.1 that the viscosity of a glass is the highest (10^{14} poises) at the lowest temperature ($500°C$ for soda lime glass, and $1100°C$ for fused silica); however, the viscosity decreases with an increase in temperature. At the highest temperature ($1600°C$), the viscosity is the lowest (1 poise for soda lime, and 10^7 poises for fused silica). The soda lime glass is readily formable into products when it has a viscosity of 10^4 poises, which is reached at a temperature around $900°C$ (working range) (see Figure 14.1). The glass is softened and undergoes steady deformation when viscosity is less than 10^8 poises at a temperature around $700°C$. The working range strongly depends on the glass composition. For example, the working range for soda lime glass is $700°C–900°C$ (approximately), which is moderately low. The working range for 96% silica glass is around $1600°C$, which is quite high (i.e., this glass is uneconomical to process for forming into objects). This is why soda lime glass is the most prevalent type of glass used for glass containers (bottles and jars), windowpanes, and other applications.

Mathematical modeling: The forming of glass into objects involves a great deal of strain rate.

The viscosity of a viscous material is generally related to strain rate by (Huda, 2017)

$$\eta = \sigma/(d\varepsilon/dt) \qquad (14.1)$$

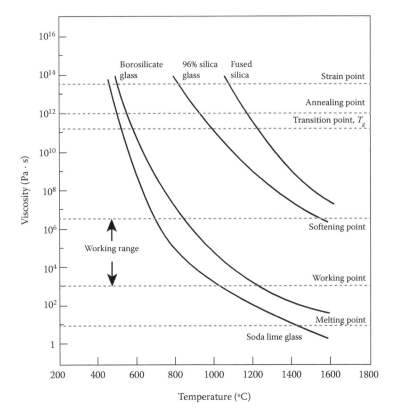

FIGURE 14.1
Viscosity versus temperature curves plot (log-linear scale) for various glasses.

where η is the viscosity of the material, poise; σ is tensile stress, Pa; and $d\varepsilon/dt$ is the strain rate.

The significances of Equation 14.1 and Figure 14.1 are illustrated in Examples 14.1 and 14.2.

14.2 Glass Production/Processing

14.2.1 Stages in Glass Production

Glass is produced by melting raw materials and often cullet in glass furnaces of different sizes employing different technologies. The raw materials include silica (SiO_2) (generally supplied as quartz sand), soda ash (Na_2CO_3), and limestone ($CaCO_3$). The stages (principal steps) in glass processing involve

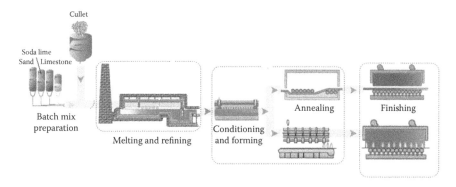

FIGURE 14.2
Stages in glass production.

(1) batch mix preparation, (2) melting and refining, (3) conditioning, (4) forming, (5) heat treatment, and (6) finishing (polishing or coating); these main
steps of glass processing are illustrated in Figure 14.2 and briefly explained
in the following paragraphs.

14.2.2 Batch Mix Preparation and Mathematical Modeling

This starting stage involves weighing fine-ground raw materials (formers,
fluxes, stabilizers, and sometimes colorants) according to the recipe required
for the final product and their subsequent mixing to achieve a homogenous
composition.

The calculations for batch mix preparation involve the determination of
gravimetric factor and the related weight calculations. The *gravimetric factor*,
G_f, is given by (Shelby, 2005)

$$G_f = M_{rm}/M_O = \text{(wt. of raw material)}/\text{(wt. of oxide)} \qquad (14.2)$$

where M_{rm} is the molar weight of raw material, g; and M_O is the molar weight
of oxide, g (see Example 14.3 and Example 14.4).

Equation 14.2 can be rearranged for weight of oxide as follows:

$$\text{Wt. of oxide} = \text{(wt. of raw material)}/G_f \qquad (14.3)$$

The significance of Equation 14.3 is illustrated in Example 14.5.

14.2.3 Melting and Refining

The melting stage involves the use of a gas-fired regenerative furnace. Here,
the *batch mix* of raw materials is automatically added at the filling end of

the furnace, the materials flow as a blanket to form molten glass at about 1550°C in the furnace. Any bubbles escape out of the molten glass so as to accomplish refining of glass.

14.2.4 Conditioning

Conditioning refers to producing a stable glass with desired glass-transition temperature, T_g (see Section 14.1). Conditioning of glass is carried out in the fore-hearths—the main channels that transport molten glass to the forming machines. The performance of the fore-hearth is rated by the range of pull rates and "gob" temperatures where the production can maintain an acceptable degree of homogeneity.

14.2.5 Forming

In order to form glass, the conditioned glass is transferred from the fore-hearth to the forming equipment at a constant rate. Depending on the type of forming process, the viscous glass stream is either continuously shaped (float-glass, fiberglass), or severed into portions of constant weight and shape ("gob"), which are delivered to a forming machine. The forming stage enables us to manufacture narrow-mouthed glass containers by use of a glass blowing machine; here, both the preliminary and final shapes are blown using compressed air. The *working temperature* range of glass forming depends on the material composition (see Example 14.2).

14.2.6 Heat Treatment

Heat treatment of glass refers to either annealing or tempering it. *Annealing* refers to heating at the annealing point until the entire glass item becomes uniformly hot and holding at the temperature steady long enough to remove all stress caused from the manufacturing process (see Example 14.5). Finally, the glass is cooled at a sufficiently slow rate so as to avoid the building up of stresses in the annealed glass. *Tempering* heat treatment of glass involves admitting pieces of glass through a tempering oven, either in a batch or continuous feed. The oven heats the glass to a temperature above 600°C (usually 620°C). The glass then undergoes a high-pressure cooling operation called *quenching*, which lasts just seconds. Here, high-pressure air blasts the surface of the glass from an array of nozzles in varying positions. *Quenching* cools the outer surfaces of the glass much more quickly than the center. As the center of the glass cools, it tends to pull back from the outer surfaces. As a result, the center remains in tension, and the outer surfaces in compression; this is why tempered glass has both strength and toughness.

14.3 Ceramics

Ceramics are nonmetallic inorganic compounds that are capable of withstanding high temperatures. Examples of ceramics include silicates (e.g., kaolinite [$Al_2Si_2O_5(OH)_4$]) and mullite ($Al_6Si_2O_{13}$), simple oxides (e.g., alumina [Al_2O_3] and zirconia [ZrO_2]), complex oxides (e.g., barium titanate [$BaTiO_3$]), carbides, nitrides, borides, and the like (Rahaman, 2003). Ceramic materials can be broadly classified into two groups: traditional ceramics and advanced ceramics.

14.4 Traditional Ceramic Processing

14.4.1 Steps in Traditional Ceramic Processing

Traditional ceramics are composed of three basic constituents: clay (~50%), silica (quartz) (~25%), and feldspar (~25%). Clay ($Al_2O_3 \cdot SiO_2 \cdot H_2O$) is the principal constituent of traditional ceramics and is extensively used in structural clay products (bricks, pipes, tiles) and white-wares (pottery, tableware, etc.). The principal steps involved in the processing of traditional ceramics include (1) powder production, (2) wet-clay forming, (3) drying, and (4) firing. These principal steps are explained in the following sections.

14.4.2 Ceramic Powder Production

First, the lumps of naturally occurring ceramic stones are broken down into small pieces by the use of crushers (see Figure 14.3a); the pieces so obtained are ground in ball mills to produce ceramic powders (see Figure 14.3b).

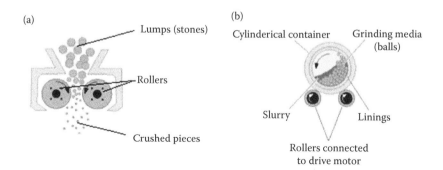

FIGURE 14.3
Crushing and grinding: (a) crushing in roller crusher and (b) grinding in ball mill.

Mathematical modeling: Figure 14.3a illustrates working principle of a roller crusher, which is capable of crushing lump down to a minimum particle size of about 2 mm. In general, a high-speed double-roller crusher has two rolls each of 1 m diameter and 65 cm length. The *peripheral speed ratio* (*i*) in a *roller crusher operation* is given by

$$i = V_{fast}/V_{slow} \qquad (14.4)$$

where V_{slow} is the peripheral speed of the slower roll, and V_{fast} is the peripheral speed of the faster roll; this mathematical relationship enables us to design a roller crusher by computing the roll diameter (see Example 14.7).

It is also possible to calculate diameter of rolls for a roller crusher when the peripheral speeds of the rolls are unknown; this calculation is accomplished by using the following formula:

$$\cos \beta = (R + r_p)/(R + r) \qquad (14.5)$$

where β is the friction angle; R is the radius of roll; r_p is the radius of particle after crushing; and r is the radius of feedstock (lump) (before rushing).

14.4.3 Wet Clay Forming

The clay powder production step is followed by powder mixing and blending with processing additives; these include lubricants, binders, plasticizers, water, and the like. There are various techniques for forming ceramic powders into desired shape, which include: (1) extrusion, (2) plastic pressing, (3) slip casting, and (4) semi-dry pressing (Mostetzky, 1978).

Extrusion: Extrusion involves forcing the ceramic paste through a die by using a ram. As a result, a long product (e.g., bar, rod, pipes, etc.) of regular cross section is produced; the long product may be cut into pieces of required length. Extruded ceramic products include furnace tubes, thermocouple components, heat exchanger tubes, and the like.

Plastic pressing: *Plastic pressing* of ceramic involves the mixing of clay with water (around 20%) to form a plastic material. The plastic clay so obtained is pressed between two mold halves (upper and lower molds). Moisture in the clay is absorbed by porous molds; then a vacuum is drawn on the back of the mold halves, resulting in the removal of moisture from the clay. Finally, the mold halves are opened and the ceramic product is removed.

Slip casting: In *slip casting*, a slurry of the wet clay powder is poured into a porous plaster mold, which absorbs the water in the ceramic slurry taking the shape of the mold cavity and forming the cake (see Figure 14.4). When a desirable thickness has been reached, the excess slurry is discharged, and the solid casting (formed part) is obtained. *Slip casting* is an ideal forming technique to produce large thin-walled hollow components.

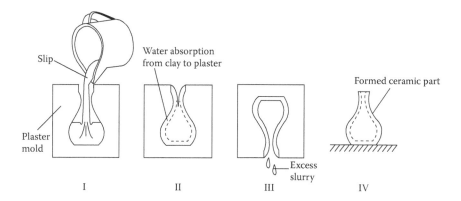

FIGURE 14.4
Slip casting; (I) Pouring slip into plaster mold, (II) water absorption and cake formation, (III) removal of excess slurry, and (IV) formed part.

Mathematical modeling of slip casting: In order to mathematically model the slip casting process, it is assumed that there is a stationary condition without any sedimentation with noncompressible cast. An increase in the cast thickness, dx, can be calculated by the following differential equation (Heimann, 2010):

$$dx/dt = (c/x) \tag{14.6}$$

where x is the cast thickness; dx is the increase in the cast thickness during the time interval dt, and c is a material constant depending on the permeability and the drain volume of the cast as well as the solid yield and the viscosity.

By rearranging the terms, Equation 14.6 takes the form

$$dx = (c/x)\, dt \tag{14.7}$$

$$\text{or} \quad \int_0^x x\, dx = \int_0^t c\, dt = c \int_0^t dt$$

$$x = \sqrt{2ct} \tag{14.8}$$

The significance of Equation 14.8 is illustrated in Example 14.9.

14.4.4 Drying and Firing of Formed Clay

Drying of formed clay enhances its strength. Dried clay product can be easily handled and stacked. *Firing* refers to the heating of dried clay in a kiln. Initially dried clay components are heated at a moderate temperature followed by firing at a high temperature in the range of 870°C–1300°C.

14.5 Advanced Ceramic Processing

Advanced or engineering ceramics exhibit diverse and often unique thermal, mechanical, chemical, or electrical/magnetic/optical properties that render them materials for applications in current and emerging technologies. Advanced ceramics find applications in electronics, communications, energy, medicine, transportation, chemical processing, and environmental sustainability. There is a wide variety of techniques for the processing of advanced ceramics; these techniques are discussed elsewhere (e.g., Biner, 1990). One of the techniques of industrial importance is the *solgel* process. In the *solgel* process, a solution of metal compounds or a suspension of very fine particles in a liquid (the *sol*) is converted into a highly viscous mass (the *gel*).

Mathematical modeling: The solgel process can be mathematically modeled by simplifying the liquid-filled pore channels in the gel as a set of parallel cylinders each of radius r. The maximum capillary stress on the solid network of the gel is given by (Rahaman, 2003)

$$\sigma_{cap} = (2\gamma_{lv} \cos \theta)/a \qquad (14.9)$$

where σ_{cap} is the maximum capillary stress, Pa; a is the radius of liquid-filled pore channels, m; γ_{lv} is the specific surface energy of the liquid-vapor interface, J/m^2; and θ is the contact angle (see Figure 13.4) (Chapter 13). The significance of Equation 14.9 is explained in Example 14.10.

14.6 Problems and Solutions—Examples in Glass and Ceramic Processing

EXAMPLE 14.1: COMPUTING VISCOSITY OF A STRAINED GLASS SAMPLE

A tensile force of 3 N is applied to a cylindrical specimen (diameter $= 4.5$ mm, length $= 120$ mm) made of borosilicate glass. It is required that the extension be less than 3 mm over a period of 5 days.

Compute the viscosity of the glass for the strain rate.

Solution

$F = 3$ N, $l = 120$ mm, $D = 4.5$ mm, $\delta l = 3$ mm, $dt = 5 \times 24 \times 3600 = 432{,}000$ s.

$$\text{Tensile stress} = \sigma = F/A = (3)/\{(\pi) \, [(4.5)^2/4]\}$$
$$= (3)/(15.9) = 0.188 \text{ MPa} = 188 \text{ kPa}$$

$$\text{Strain} = d\varepsilon = \delta l/l = 3/120 = 0.025$$

$$\text{Strain rate} = d\varepsilon/dt = 0.025/432000 = 5.78 \times 10^{-8} \text{ s}^{-1}$$

By using Equation 14.1,

$$\eta = \sigma/(d\varepsilon/dt) = (188{,}000)/(5.78 \times 10^{-8})$$
$$= 32{,}526 \times 10^{8} = 3.2 \times 10^{12} \text{ poise}$$

The viscosity of the strained borosilicate glass $= 3.2 \times 10^{12}$ poises.

EXAMPLE 14.2: DETERMINING THE MAXIMUM AND WORKING TEMPERATURES FOR A STRAINED GLASS

Refer to the data in Example 14.1. (a) Determine the maximum temperature to which the borosilicate glass specimen should be heated to achieve the desired viscosity. (b) Is this temperature suitable for forming of the glass? Explain! (c) Determine a suitable working temperature range for the glass.

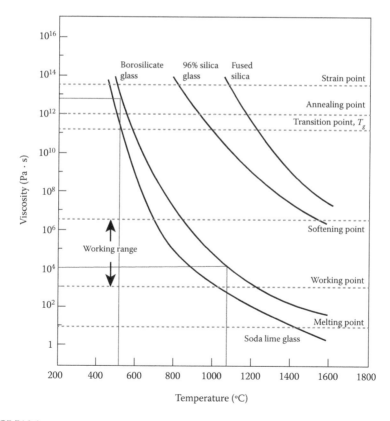

FIGURE E14.1
Drawn lines in Figure 14.1 to obtain the temperature for $\eta = 3.2 \times 10^{12}$ poise and to determine a working temperature for borosilicate glass.

Solution

(a) By reference to Figure 14.1, we draw lines for borosilicate glass (Viscosity $= 3.2 \times 10^{12}$ poise) in the graphical plot to obtain Figure E14.1.

In Figure E14.1, we can see that the viscosity $= 3.2 \times 10^{12}$ poise for borosilicate glass corresponds to the maximum temperature close to 550°C.

(b) The temperature (550°C) is not suitable for forming of the glass because it does not correspond to a viscosity within the working range; it is in the range between *annealing point* and *strain point*.

(c) In order to obtain a temperature that corresponds to a viscosity within the working range, we take viscosity $= 10^4$ poises and draw lines for borosilicate glass as shown in Figure E14.1. Thus, the working temperature should be around 1050°C.

EXAMPLE 14.3: CALCULATING THE GRAVIMETRIC FACTOR OF LIMESTONE

Limestone ($CaCO_3$) is one of the raw materials in glass production. Calculate the *gravimetric factor* of limestone.

Solution

During melting of the raw materials, limestone ($CaCO_3$) decomposes into calcium oxide (CaO) and carbon dioxide gas.

Gram molar weight of raw material, $CaCO_3 = 40 + 12 + 3(16) = 100\,g$

$$\text{Gram molar weight of oxide, } CaO = 40 + 16 = 56\,g$$

By using Equation 14.2,

$$G_f = M_{rm}/M_O = 100/56 = 1.78$$

The *gravimetric factor* of limestone $= 1.78$.

EXAMPLE 14.4: CALCULATING REQUIRED WEIGHT OF LIMESTONE FOR PRODUCING AN OXIDE

Calculate the weight of limestone that is required for producing 130 grams CaO.

Solution

By reference to Example 14.3, G_f of limestone $= 1.78$.

By using Equation 14.2,

$$G_f = (\text{wt. of raw material})/(\text{wt. of oxide})$$

$$1.78 = (\text{wt. of raw material})/(130\,g)$$

Weight of raw material $= 1.78 \times 130 = 232$ g

The required weight of limestone $= 232$ grams.

EXAMPLE 14.5: CALCULATING THE COMPOSITION OF GLASS PRODUCED FROM KNOWN BATCH MIX

A batch mix comprising 150 grams of K_2CO_3, 90 grams of limestone ($CaCO_3$), and 500 grams of silica (SiO_2) was melted for glass production. Calculated the wt.% composition of the glass.

Solution

Wt. $K_2CO_3 = 150$ g, wt. $CaCO_3 = 90$ g, wt. $SiO_2 = 500$ g.

First, we calculate the gravimetric factors of each raw material as illustrated in Example 14.3:

$$G_f \text{ of } K_2CO_3 = 1.47$$

$$G_f \text{ of } CaCO_3 = 1.78$$

$$G_f \text{ of } SiO_2 = 1.0/1 = 1$$

Next, we calculate the amount of each oxide by using Equation 14.3 as follows:

$$\text{Wt. } K_2O = (\text{wt. } K_2CO_3)/(G_f \text{ of } K_2CO_3) = 150/1.47 = 102 \text{ g}$$

$$\text{Wt. } CaO = (\text{wt. } CaCO_3)/(G_f \text{ of } CaCO_3) = 90/1.78 = 50.5 \text{ g}$$

$$\text{Wt. } SiO_2 = (\text{wt. } SiO_2)/(G_f \text{ of } SiO_2) = 500/1 = 500 \text{ g}$$

$$\text{Total weight of oxides} = 102 + 50.5 + 500 = 652.5 \text{ g}$$

$$\text{Wt\% } K_2O = [(102)/(652.5)] \times 100 = 15.6$$

$$\text{Wt\% } CaO = [(50.5)/(652.5)] \times 100 = 7.7$$

$$\text{Wt\% } SiO_2 = [(500)/(652.5)] \times 100 = 76.6$$

The composition of glass: $K_2O = 15.6$ wt.%; $CaO = 7.7$ wt.%; $SiO_2 = 76.6$ wt.%.

In order to determine the batch mix composition required for a specific glass composition, the above calculations should be reversed.

EXAMPLE 14.6: DETERMINING AND COMPARING THE ANNEALING POINTS OF VARIOUS GLASSES

Determine the annealing points of the following glasses: soda lime glass, borosilicate glass, 96% silica glass, and fused silica glass. Which glass is the most economical to anneal?

Solution

By reference to Figure 14.1, we obtain the following:

Annealing point of soda lime glass $= 520°C$
Annealing point of borosilicate glass $= 580°C$
Annealing point of 96% silica glass $= 940°C$
Annealing point of fused silica $= 1100°C$

Since $520°C$ is the lowest annealing point, the annealing of soda lime glass is the most economical.

EXAMPLE 14.7: DESIGNING A ROLLER CRUSHER BY CALCULATING ROLL DIAMETER WHEN THE PERIPHERAL SPEEDS OF ROLLS ARE KNOWN

Design a roller crusher by calculating its (a) peripheral speed ratio, (b) diameter of the slower roll, and (c) diameter of the faster roll. The given data are as follows. The peripheral speed of the slower roll $= 11.7\,m/s$; the peripheral speed of the faster roll $= 17.25\,m/s$. The rotational speed of the slower roll $= 100\,rpm$; the rotational speed of the faster roll $= 150\,rpm$.

Solution

$V_{slow} = 11.7\,m/s$; $V_{fast} = 17.25\,m/s$; $\omega_s = 100\,rpm$; $\omega_f = 150\,rpm$.

(a) By using Equation 14.4,

$$i = V_{fast}/V_{slow} = (17.25)/(11.7) = 1.5$$

The peripheral speed ratio $= 1.5$.

(b) $\omega_s = 100\,rev/min = (100 \times 2\pi)\,rad./60\,s = 628.3/60 = 10.5\,rad./s$

$$V_{slow} = r\,\omega_s$$

$$11.7 = r\,(10.5)$$

$$r = 11.7/10.5 = 1.11\,m, \text{ Diameter} = 2r = 2 \times 1.11 = 2.2\,m$$

The diameter of the slower roll $= 2.2\,m$.

(c) $\omega_f = 150\,rev/min = (150 \times 2\pi)\,rad./60\,s = 15.7\,rad./s$

$$V_{fast} = r\omega_f,$$

$$17.25 = r\,(15.7)$$

$$r = 17.25/15.7 = 1.1\,m, \text{ Diameter} = 2r = 2 \times 1.1\,m = 2.2\,m$$

The diameter of faster roll $= 2.2\,m$.

Hence, the roller crusher should be designed so that each roll has a 2.2 m diameter; this design will have a peripheral speed ratio of 1.5.

EXAMPLE 14.8: CALCULATING THE DIAMETER OF ROLL FOR A ROLLER CRUSHER WHEN THE PERIPHERAL SPEED OF ROLL IS UNKNOWN

A pair of rolls of a roller crusher is to take a stock feed equivalent to spheres of 3.5 cm in radius and crush them to spherical particles having 1.5 cm radius. Calculate the diameter of each roll, if the coefficient of friction is 0.3.

Solution

Coefficient of friction $= \mu = 0.3$, $r = 3.5$ cm, $r_p = 1.5$ cm, $R = ?$

The coefficient of friction (μ) is related to the friction angle (β) by (see Chapter 8)

$$\mu = \tan \beta$$

$$0.3 = \tan \beta$$

$$\beta = 16.7°$$

By using Equation 14.5,

$$\cos \beta = (R + r_p)/(R + r)$$

$$\cos 16.7° = (R + 1.5)/(R + 3.5)$$

$$0.96 = (R + 1.5)/(R + 3.5)$$

$$R + 1.5 = 0.96\,R + 3.35$$

$$R = 46.2 \text{ cm}$$

$$\text{Dia.} = 2R = 2 \times 46.2 = 92 \text{ cm}$$

The diameter of each roll $= 92$ cm.

EXAMPLE 14.9: CALCULATING THE CONSTANT c WHEN SLIP CAST THICKNESS AND CASTING TIME ARE GIVEN

A ceramic bowl was formed by slip casting in 10 minutes. The wall thickness of the slip cast bowl is 2 mm. Calculate the material constant c for the Equation 14.6.

Solution

$t = 10$ min, $x = 2$ mm, $c = ?$

By using Equation 14.8,

$$x = \sqrt{2ct}$$

$$2\,\text{mm} = \sqrt{2c\,(10\,\text{min})}$$

$$c = 0.2\,\text{mm}^2/\text{min}$$

EXAMPLE 14.10: COMPUTING THE MAXIMUM CAPILLARY STRESS FOR AN ALKOXIDE-DERIVED GEL

The specific surface energy of the liquid-vapor interface for an alkoxide-derived gel is 0.06 J/m².

The contact angle is 20°, and the radius of liquid-filled pore channels is 6 nm. Compute the maximum capillary stress.

Solution

$\gamma_{ly} = 0.06\,\text{J/m}^2$, $a = 6\,\text{nm} = 6 \times 10^{-9}\,\text{m}$, $\theta = 20°$, $\sigma_{cap} = ?$

By using Equation 14.9,

$$\sigma_{cap} = (2\gamma_{ly}\cos\theta)/a = (2 \times 0.06 \times \cos 20°)/(6 \times 10^{-9})$$
$$= 0.018 \times 10^3 \times 10^6\,\text{Pa} = 18\,\text{MPa}$$

The maximum capillary stress = 18 MPa.

Questions and Problems

14.1 Draw a labeled sketch showing the stages in glass production.

14.2 (a) Why is soda lime glass the most prevalent type of glass in the industrial world? (b) List at least three application areas of glasses.

14.3 Differentiate between the following: (a) glass-transition temperature and melting temperature; (b) glass annealing and glass tempering.

14.4 Why is tempered glass stronger and tougher as compared to annealed glass?

14.5 (a) What are the constituents of traditional ceramics? (b) List the main steps in traditional ceramic production.

14.6 Draw labeled sketches showing the following operations in ceramic processing: (a) *crushing* in a roller crusher and (b) *grinding* in a ball mill.

14.7 Explain *slip casting* of a clay component with the aid of sketches.

P14.8 A tensile force of 4 N is applied to a cylindrical specimen (diameter = 5 mm, length = 140 mm) of borosilicate glass. It is required that the extension be less than 3.3 mm over a period of 5 days. Compute the viscosity of the glass for the strain rate.

P14.9 By reference to Figure 14.1, determine the viscosity of 96% silica glass at a temperature of 1200°C. Is this temperature suitable for forming of 96% silica glass? Justify!

P14.10 A batch mix comprising 200 grams of K_2CO_3, 120 grams of limestone ($CaCO_3$), and 700 grams of silica (SiO_2) was melted for glass production. Calculated the wt.% composition of the glass.

P14.11 Potassium carbonate (K_2CO_3) is an important raw material in glass processing. Calculate the gravimetric factor of potassium carbonate.

P14.12 A ceramic bowl was formed by slip casting in 13 minutes. The wall thickness of the slip cast bowl is 2.1 mm. Calculate the material constant c for the Equation 14.6.

P14.13 Refer to Figure 14.1. Which glass is the least economical to anneal? Justify your selection.

14.14 Design a roller crusher by specifying the peripheral speeds of its slower and faster rolls if the *peripheral speed ratio* is 1.7.

P14.15 A pair of rolls of a roller crusher is to take a stock feed equivalent to spheres of 3.8 cm in diameter and crush them to spherical particles having 1.7 cm diameter. Calculate the diameter of each roll, if the coefficient of friction is 0.28.

References

Biner, J.G.P. 1990. *Advanced Ceramic Processing and Technology*. New York, NY: Elsevier Science.

Heimann, R.B. 2010 *Classic and Advanced Ceramics: From Fundamentals to Applications*. New York, NY: Wiley.

Huda, Z. 2017. *Materials Processing for Engineering Manufacture*. Zurich, Switzerland: Trans Tech.

Mostetzky, H. 1978. Shaping in the ceramic. *Handbook of Ceramics*, Group I D1, Schmid Freiburg, Germany: Schmid GmbH, 1–12.

Ojovan, M.I. 2008. Viscosity and Glass Transition in Amorphous Oxides. *Advances in Condensed Matter Physics*, Article ID 817829. 2008, 1–23.

Rahaman, M.N. 2003. *Ceramic Processing and Sintering*, 2nd Edition. Boca Raton, FL: CRC Press.

Shelby, J.E. 2005. *Introduction to Glass Science and Technology*. Cambridge, UK: Royal Society of Chemistry.

15

Composite Processing

15.1 Composites and Their Processing

15.1.1 Composites and Their Classification

A material that is composed of two or more distinct phases (matrix and dispersed) and having bulk properties significantly different from those of any of the constituents is called a *composite*. Examples of composite components include fiberglass tanks, helmets, sport shoes, sailboat masts, light poles, compressed natural gas (CNG) tanks, rescue air tanks, and the like. In particular, carbon-fiber reinforced polymer (CFRP) composites are increasingly being used in the wings and fuselage of aerospace structures (Huda and Edi, 2013).

Based on the matrix materials, composites may be classified into four groups: (1) polymer matrix composites (PMCs), (2) metal matrix composites (MMCs), (3) ceramic matrix composites (CMC), and (4) carbon matrix composites. Based on reinforcement form, composites may be divided into three categories: (1) particle reinforced composites, (2) fiber reinforced (FR) composites (short fibers and continuous fibers), and (3) laminates.

15.1.2 Composite Processing—Principal Steps and Types

15.1.2.1 Principal Steps in Composite Processing

Composite processing generally involves four main steps: (1) impregnation, (2) laying up, (3) consolidation, and (4) solidification (Mazumdar, 2002). These steps are explained as follows.

Step 1: Impregnation/Wetting—*Impregnation* is the first step in composite processing; it involves the mixing of resins and fibers so as to ensure that the resin flows entirely around all fibers. The factors controlling the impregnation process include viscosity, surface tension, capillary action, and the like. Thermoset resins (matrix) are easier to wet (impregnate) owing to their lower viscosities. Thermoplastic resins require the application of greater pressure for impregnation.

Step 2: Lay Up—This step involves the formation of laminates by placing fiber/resin mixtures (or prepregs) at the right places and angles so as to ensure the right fiber orientation as per fiber architecture dictated by the composite design. In general, layers of resin and fibers are laid one over the other to build up the desired composite thickness.

Step 3: Consolidation—*Consolidation* refers to the development of an effective contact between each layer of the laminate by the application of pressure. Consolidation ensures that the composite part is free from voids and dry spots.

Step 4: Solidification—Finally, the consolidated laminate is allowed to solidify. The solidification time varies in the range from 1 min (for thermoplastic matrix composites) to 2 hours (for thermoset matrix composites).

15.1.2.2 Types of Composite Manufacturing Processes

Among the various categories of composites mentioned in the preceding section, PMC products dominate over all other categories. Accordingly, there are two main types of processing based on PMCs: thermoset matrix composite processing and thermoplastic matrix composite processing; the former is the most generally used type of composite processing. Since thermoset matrix composite products constitute more than three-fourths of all manufactured composite parts, it is reasonable to focus on the thermoset composite processing techniques, which include (1) hand lay-up process, (2) prepreg lay-up process, (3) spray up process, (4) pultrusion process, (5) filament winding process, and (6) resin transfer molding process (Huda, 2017). These composite processing techniques are discussed in Section 15.3.

15.2 Effects of Fiber Orientation on Composite Processing and Properties

15.2.1 The Industrial Importance of Fiber Orientation in Composites

The processing technique for manufacturing FR composites strongly depends on the fiber orientation. It has been experimentally shown that unidirectional alignment of short fibers achieved by an extrusion process increases the tensile strength and elastic modulus of the composite (measured along the axis of fibers alignment) by more than double as compared to randomly oriented fiber composite (Joseph et al., 1993). For this reason, fibrous composite manufacturers often rotate layers of continuous fibers so as to align the fibers in a direction parallel to the applied stress (see Figure 15.1[a]). The strong dependence of fiber orientation on composite's properties can also be justified by the mathematical models developed in the following section.

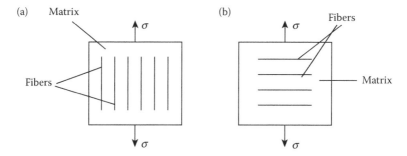

FIGURE 15.1
Loading conditions in a FR composite: (a) iso-strain condition and (b) iso-stress condition.

15.2.2 Iso-Strain and Iso-Stress Loading Conditions in Fiber Reinforced Composites

Figure 15.1 illustrates two types of loading conditions in FR composites. In the iso-strain condition, the continuous fibers are aligned parallel to the direction of stress so that phases (matrix and fibers) experience the same strain (see Figure 15.1[a]). In the iso-stress conditions, the continuous fibers are aligned perpendicular to the stress direction so that each phase experiences the same stress (see Figure 15.1[b]).

Iso-strain condition: For the iso-strain condition (Figure 15.1[a]), the stiffness or elastic modulus of the entire composite when measured along the length of the fiber is governed by the rule of mixtures and is given by

$$E_c = E_f \cdot V_f + E_m \cdot V_m \tag{15.1}$$

where E_c is the stiffness or the elastic modulus of the entire composite measured along the length of the fiber, E_f is the stiffness of the fiber along the length of the fiber, V_f is the volume fraction of the fibers in the composite, E_m is the stiffness of the matrix, and V_m is the volume fraction of the matrix. The volume fraction of matrix is related to volume fraction of fiber by

$$V_m + V_f = 1 \tag{15.2}$$

By the application of the rule of mixture, the tensile strength of a composite can be related to the strengths of the fibers and the matrix as follows:

$$(S_{ut})_c = (S_{ut})_f V_f + (S_{ut})_m V_m \tag{15.3}$$

where $(S_{ut})_c$ is the tensile strength of the composite; $(S_{ut})_f$ is the tensile strength of the fiber measured along the length of the fiber; and $(S_{ut})_m$ is the tensile strength of the matrix.

The significance of Equations 15.1 through 15.3 is illustrated in Example 15.1 and Example 15.2.

Iso-stress condition: For the iso-stress condition (Figure 15.1[b]), the stiffness or elastic modulus of the entire composite is given by

$$E_c = (E_f E_m)/[(E_f V_m) + (E_m V_f)] \qquad (15.4)$$

where the symbols E_f, V_m, E_m, and V_f have their usual meanings (see Example 15.3).

The solutions of Example 15.1 and Example 15.3 show that the stiffness of the glass fiber reinforced polymer (GFRP) composite under iso-strain condition is 44.7 GPa, whereas the stiffness of the composite under iso-stress condition is 7.28 GPa. It means that the stiffness of the composite under iso-strain condition is six times that of the same composite when loaded under the iso-stress condition. In Section 15.2.1, the industrial importance of fiber orientation in composites was emphasized. The stiffness data thus justify why fibrous composite manufacturers often rotate layers of continuous fibers so as to align them in a direction parallel to the stress (see Figure 15.1[a]).

15.3 Metal Matrix Composite Processing

Manufacturing process: MMCs are made of a continuous metallic matrix and one or more discontinuous reinforcing phases (e.g., fibers, whiskers, or particles). In a broader context, the MMC matrix material may be a metal, an alloy, or an intermetallic compound (see Example 15.4 and Example 15.5).

There are various techniques for manufacturing MMCs. One of the techniques is the liquid phase fabrication, which involves incorporation of a dispersed phase into a molten matrix metal, followed by its solidification. In order to achieve excellent mechanical properties of the composite, it is important to ensure an appropriate fiber orientation as well as a good interfacial bonding (wetting) between the dispersed phase and the liquid matrix. The good wetting is generally achieved by coating the dispersed phase particles (fibers). For example, an aluminum alloy (metal matrix)–continuous boron (fibers) MC can be produced by using a tungsten (W) wire core that is coated with boron (B) to make a fiber that is aligned parallel to the design stress (iso-strain condition) in the metal matrix.

Mathematical modeling: The application of the rule of mixtures (RoM) to a FR composite is explained in the preceding section. The RoM, when applied to the aluminum alloy–boron (with tungsten core) MMC in the iso-strain condition, enables us to express the stiffness (modulus of elasticity) of the MMC as follows:

$$E_{MMC} = E_{Al} V_{Al} + E_W V_W + E_B V_B \qquad (15.5)$$

where E_{MMC} is the stiffness of the MMC, GPa; E_{Al} is the stiffness of the aluminum matrix, GPa; V_{Al} is the volume fraction of the aluminum alloy; E_W is the stiffness of tungsten wire, GPa; V_W is the volume fraction of the tungsten wire; E_B is the stiffness of the boron fiber, GPa; and V_B is the volume fraction of the boron fiber (see Example 15.7).

If the volume fraction of W wire coated with boron fiber is known, the volume fraction of *Al* matrix can be determined by

$$V_{Al} = 1 - V_{W+B} \tag{15.6}$$

where V_{W+B} is the volume fraction of the boron fiber (tungsten wire coated with boron).

The volume fraction of tungsten wire (V_W) can be calculated by

$$V_W = (A_W/A_{W+B}) \cdot V_{W+B} \tag{15.7}$$

where A_W is the cross sectional (c.s.) area of the tungsten wire; and A_{W+B} is the c.s. area of the boron fiber.

The volume fraction of boron (V_B) can be computed by

$$V_B = [(A_{W+B} - A_W)/A_{W+B}] \cdot V_{W+B} \tag{15.8}$$

where the symbols have their usual meanings (see Example 15.6).

15.4 Thermoset Composite Manufacturing Techniques

In Section 15.1.2.2, various types of composite manufacturing processes were listed down. These processes are discussed with particular reference to FR thermoset composite processing in the following sections.

15.4.1 Hand Lay-Up Process

In the hand lay-up process, layers of composite fiber are built up in a sequenced lay-up using a matrix of resin and hardener to form a laminate stack. The laminate stack is then *"cured."* The simplest curing method involves allowing *cure* to occur at room temperature. However, heat and/or pressure may be applied to accelerate *curing*.

15.4.2 Prepreg Lay-Up Process

The prepreg lay-up process involves processing of graphite/epoxy prepregs. Prior to processing, the prepregs are refrigerated and then slowly brought to room temperature. The prepreg is then cut to the desired length and shape in a neat and clean environment that is humidity and temperature controlled. It is important to ensure the desired fiber orientation during cutting of plies (see Section 15.2). A computer-controlled machine is capable of cutting several layers of prepregs stacked together at one time. In order to fabricate a part, the prepregs are laid layer by layer on top of an open mold, which has been coated with a release agent for easy removal of the part. The prepregs are then vacuum bagged and cured (heated) at a temperature in the range of 130°C–180°C, which allows the epoxy resin to initially reflow and eventually to cure. Additional pressure (~5 atm) for the molding is usually provided by an autoclave (a pressurized oven).

15.4.3 Pultrusion

Process: *Pultrusion* is generally used to manufacture thermoset composite components that have continuous lengths and constant cross-sectional shapes (e.g., tubes, bars, beams, etc.). *Pultrusion* involves pulling reinforcing fiber from a set of fiber reels and through a heated resin bath. The principal reinforcements are glass, carbon, and aramid fibers that are normally added in a concentration in the range of 40–70 vol.%. Commonly used resin is the epoxy resin; however, other matrix materials (polyesters, vinyl esters, etc.) are also used (Callister, 2007).

The reinforcing fibers, after passing through the hot resin, are formed into specific shapes as passing through one or more pre-formers (see Figure 15.2). The material then moves through a forming and curing die, where it is formed to its net shape and is heated (*cured*). On cooling, the resulting profile is cut to desired length by use of a cut-off saw.

Mathematical modeling: The *pultrusion* process can be mathematically modeled by attaining the optimum operating condition in the forming and curing die during the composite processing. In particular, the *curing stage* is a polymeric exothermic reaction. The important curing process parameters include curing ratio (α), curing speed, rate constant (k), and thermal activation energy (E). The *curing ratio* is a function of two variables: temperature (T) and time (t).

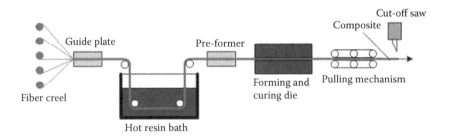

FIGURE 15.2
Pultrusion process for composite manufacture.

The curing ratio can be computed in two steps by the following Arrhenius equation (Hardis, 2012):

$$k = A_0 \exp(-E/RT) \qquad (15.9)$$

where k is the rate constant; A_0 is the pre-exponential constant; E is the thermal activation energy; R is the universal gas constant; and T is the temperature, K.

Once k is known, the curing ratio (α) can be calculated as follows (Chachad et al., 1995):

$$\alpha = 1 - [kt(n-1) + 1]^{1/(1-n)} \qquad (15.10)$$

where t is the curing time, and n is the order of the exothermic reaction (see Examples 15.8).

The curing kinetics has been quantitatively studied for the pultrusion process, and a useful mathematical model has experimentally been developed. The researchers have expressed the effective curing rate by using heat of reaction of 280 J/g, as follows:

$$k_{eff} = 0.51 \exp\{(-57.7 \times 106)\,(1/R)\,[(1/T) - (1/373)]\}/\text{min} \qquad (15.11)$$

where k_{eff} is the effective curing rate dependent on temperature, catalyst concentration, and initial iso-cyanate concentration (see Example 15.9).

15.4.4 Filament Winding Process

The filament winding process involves accurate positioning of continuous reinforcing fibers in a predetermined pattern to form a hollow cylindrical shape. First, the fibers are fed through a resin bath, and the composite is continuously wound onto a mandrel. Specific mechanical properties can be achieved through engineered fiber orientation and thickness as well as through proper selection of reinforcement and resin.

15.4.5 Resin Transfer Molding (RTM) Process

The RTM or liquid molding process uses extremely low-viscosity resin and fiberglass reinforcement to produce composite parts. First, a two-part, matched, closed metallic mold is prepared. Second, a fiber pack (a preform) is placed into the mold and the mold is closed. Third, measured quantities of resin and catalyst are mixed in dispensing equipment. Then, the mixture is injected (pumped) into the mold under moderate pressure through injection ports, following predesigned paths through the preform. Sometimes both mold and resin are heated for particular applications.

15.5 Particulate Composite Processing

Particulate composites are the composites that consist of either coarse or tiny particles embedded in another so as to produce an unusual combination of properties. Examples of particulate composites include concrete, grinding wheels, cemented carbide cutting tools, electrical contacts, and the like.

15.5.1 Concrete Manufacturing Process

Concrete is a composite made by using cement, aggregate (sand or gravel), water, and additives. The manufacture of concrete involves mixing and blending of cement and aggregates, admixtures (chemical additives), any necessary fibers, and water. A ready-mix concrete plant consists of silos that contain cement, sand, gravel, and storage tanks of additives such as plasticizers, as well as a mixer to blend the components of concrete. For normal concrete, the ranges by absolute volume of the major components are presented in Table 15.1.

15.5.2 Silver-Tungsten Electrical Composite Manufacturing Process

The electrical contacts find applications in switches and relays. They are generally made of a silver-tungsten particulate composite. The silver-tungsten composite is manufactured by infiltration of liquid silver in a powder-formed tungsten compact, thereby imparting a good combination of wear resistance and electrical conductivity. The steps in manufacturing the silver-tungsten composite are illustrated in Figure 15.3.

The first step involves the compaction of tungsten (W) powder by using a press—a typical technique of powder metallurgy (P/M) (see Figure 15.3[a]). The green W compact has low density due to the high volume fraction of porosity (Figure 15.3[b]). Then, sintering of W compact results in a decrease of voids volume fraction (see Figure 15.3[c]). Finally, the interconnected voids in the compact are filled by vacuum infiltration of silver to form the composite (see Figure 15.3[d]).

TABLE 15.1

Major Constituents of Concrete (vol.%)

Number	Constituent	Vol.% range
1	Cement	7–15
2	Fine aggregate	25–30
3	Coarse aggregate	31–51
4	Water	16–21

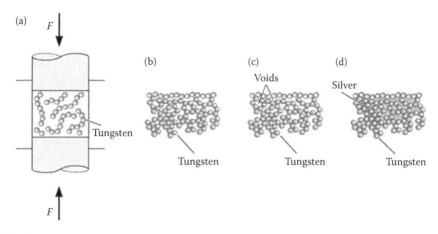

FIGURE 15.3
Manufacturing of silver-tungsten (particulate) composite: (a) compaction of tungsten (W) particulates, (b) green W compact, (c) sintered W compact, and (d) silver infiltrated W compact.

15.5.3 Mathematical Modeling of Particulate Composites

The mechanical behavior and properties of a particulate composite are largely controlled by the relative amounts and properties of the individual constituents of a particulate composite. For example, the properties of a silver-tungsten composite strongly depend on the volume fractions of tungsten (W) particulates and the infiltrated silver (Ag) in the composite. The RoM can be applied to a particulate composite to express the composite's density as follows:

$$\rho_c = V_1\rho_1 + V_2\rho_2 + V_3\rho_3 + \cdots + V_n\rho_n \tag{15.12}$$

where ρ_c is the density of the particulate composite; $V_1, V_2, V_3, \ldots, V_n$ are the volume fractions of each constituent in the composite; and $\rho_1, \rho_2, \rho_3, \ldots, \rho_n$ are the densities of each constituent in the particulate composite. The sum of all volume fractions is equal to unity, i.e.,

$$V_1 + V_2 + V_3 + \cdots + V_n = 1 \tag{15.13}$$

It must be noted that RoM (Equation 15.12) is applicable to the particulate composite, provided the dispersed particles are well connected to the continuous phase (matrix) (see Example 15.11).

15.5.4 Dispersion Strengthened Nanocomposite Processing

Dispersion strengthened (DS) composites possess excellent high-temperature strength suitable for application in hot sections of gas-turbine engines' components. An important DS composite is thoria-dispersed nickel (TD-nickel), which is produced by using P/M manufacturing cycle. First, measured amounts of thorium (Th) and nickel (Ni) powders are ball milled. Second, the powder blend is compacted at high pressure in a press. Finally, the compact is sintered (heated) in air so as to allow the diffusion of atmospheric oxygen into the compact; the oxygen reacts with thorium (Th) metal to form a fine dispersion of thoria (ThO_2). This processing results in a fine dispersion of hard nanosized thoria particles in a relatively ductile nickel matrix (see Example 15.12).

15.6 Problems and Solutions—Examples in Composite Processing

EXAMPLE 15.1: CALCULATING THE ELASTIC MODULUS OF A GFRP COMPOSITE UNDER ISO-STRAIN CONDITION

The elastic modulus of a glass fiber is 72.4 GPa, and its tensile strength is 2.4 GPa. A GFRP has been produced by reinforcing the glass fibers into the polymer matrix. The GFRP composite has 60% by volume of continuous glass fibers and a hardened epoxy resin having elastic modulus and tensile strength of 3.1 GPa and 62 MPa, respectively. Determine the elastic modulus of the GFRP composite under an iso-strain condition.

Solution

In this problem, the fiber is glass fiber and the matrix is epoxy resin:

$$E_f = 72.4 \text{ GPa}, \ V_f = 60\% = 0.60, \ E_m = 3.1 \text{ GPa}$$

By using Equation 15.2,

$$V_m = 1 - V_f = 1 - 0.60 = 0.40$$

By using Equation 15.1,

$$E_c = E_f V_f + E_m V_m = (72.4 \times 0.6) + (3.1 \times 0.4) = 43.44 + 1.24 = 44.7 \text{ GPa}$$

The elastic modulus of the GFRP composite under iso-strain condition = 44.7 GPa.

EXAMPLE 15.2: COMPUTING THE TENSILE STRENGTH OF A GFRP COMPOSITE UNDER ISO-STRAIN CONDITION

By using the data in Example 15.1, calculate the tensile strength of the GFRP composite.

Solution

$V_m = 0.40$, $V_f = 0.60$, $(S_{ut})_f = 2.4\,GPa = 2400\,MPa$, $(S_{ut})_m = 62\,MPa$, $(S_{ut})_c = ?$

By using Equation 15.3,

$$(S_{ut})_c = (S_{ut})_f\,V_f + (S_{ut})_m\,V_m = (2400 \times 0.6) + (62 \times 0.4) = 1464.8\,MPa$$

The tensile strength of the GFRP composite = 1464.8 MPa.

EXAMPLE 15.3: CALCULATING THE ELASTIC MODULUS OF A GFRP COMPOSITE UNDER ISO-STRESS CONDITION

By using the data in Example 15.1, determine the stiffness of the GFRP composite under an iso-stress condition.

Solution

$E_f = 72.4\,GPa$, $V_f = 0.60$, $E_m = 3.1\,GPa$, $V_m = 0.40$.

By using Equation 15.4,

$$E_c = (E_f\,E_m)/[(E_f\,V_m) + (E_m V_f)] = (72.4 \times 3.1)/[(72.4 \times 0.4)$$
$$+ (3.1 \times 0.6)] = 7.28\,GPa$$

The stiffness of the GFRP composite under iso-stress condition = 7.28 GPa.

EXAMPLE 15.4: CALCULATING THE STIFFNESS OF AN MMC

An MMC is made of an aerospace aluminum alloy matrix and 45 vol.% alumina (Al_2O_3) continuous fibers oriented in a direction parallel to the design stress. Calculate the elastic modulus of the MMC in the iso-strain condition. The stiffness data are $E_{alumina} = 395\,GPa$ and $E_{Al\,alloy} = 70\,GPa$.

Solution

$V_f = 45\% = 0.45$, $E_f = 395\,GPa$, $E_m = 70\,GPa$, $E_c = ?$

By using Equation 15.2,

$$V_m = 1 - V_f = 1 - 0.45 = 0.55$$

By using Equation 15.1,

$$E_c = E_f V_f + E_m V_m = (395 \times 0.45) + (70 \times 0.55)$$
$$= 177.75 + 38.5 = 216.25\,GPa$$

The elastic modulus of the MMC = 216 GPa.

EXAMPLE 15.5: COMPUTING VOLUME FRACTION OF FIBER IN AN MMC

Titanium aluminide (Ti_3Al) is an intermetallic compound. A new MMC is made of titanium aluminide and continuous silicon carbide (SiC) fibers oriented in the iso-strain condition. What must be the volume fraction of SiC if the elastic modulus of the composite is 240 GPa? The stiffness data are given as follows: $E_{Ti_3Al} = 145$ GPa, and $E_{SiC} = 390$ GPa.

Solution

$E_c = 240$ GPa, $E_m = 145$ GPa, $E_f = 390$ GPa, $V_f = ?$

By substituting $[V_m = 1 - V_f]$ from Equation 15.2 into Equation 15.1, we obtain

$$E_c = E_f \cdot V_f + E_m \cdot Vm = E_f \cdot V_f + E_m(1 - V_f)$$

$$240 = [390V_f + 145(1 - V_f)]$$

$$390V_f + 145 - 145V_f = 240$$

$$V_f = 95/245 = 0.38$$

The volume fraction of the silicon carbide fibers in the MMC = 0.38.

EXAMPLE 15.6: CALCULATING VOLUME FRACTION OF MATRIX AND FIBER IN AN MMC

An MMC is made of an aerospace Al alloy matrix and continuous boron fibers. The boron fibers are produced using a 12 μm diameter tungsten-wire core that is coated with boron. The final diameter of the boron fiber is 108 μm. The volume fraction of the boron fiber is 52%. The boron fibers are oriented in the iso-strain condition in the Al alloy matrix. Calculate the volume fractions of (a) aluminum alloy matrix, (b) tungsten wire, and (c) boron.

Solution

$V_{W+B} = 52\% = 0.52$, $A_W = (\pi/4) (12\,\mu m)^2$, $A_{B+W} = (\pi/4) (108\,\mu m)^2$

By using Equation 15.5,

$$V_{Al} = 1 - V_{W+B} = 1 - 0.52 = 0.48$$

By using Equation 15.7,

$$V_W = (A_W/A_{W+B}) \cdot V_{W+B} = \{[(\pi/4) (12\,\mu m)^2]/[(\pi/4) (108\,\mu m)^2]\} \times 0.52$$

$$V_W = [(12\,\mu m)^2/(108\,\mu m)^2] \times 0.52 = (12^2/108^2) \times 0.52 = 0.006$$

By using Equation 15.8,

$$V_B = [(A_{W+B} - A_W)/A_{W+B}] \cdot V_{W+B}$$

$$V_B = \{\{[(\pi/4)\,(108\,\mu m)^2] - [(\pi/4)\,(12\,\mu m)^2]\}/[(\pi/4)\,(108\,\mu m)^2]\} \times 0.52$$

$$V_B = [(108^2 - 12^2)/108^2] \times 0.52 = [(11664 - 144)/11664] \times 0.52 = 0.514$$

$V_{Al} = 0.48,\ V_W = 0.006,\ V_B = 0.514$

Verification: $V_{Al} + V_W + V_B = 0.48 + 0.006 + 0.514 = 0.999 \cong 1$.

Hence, the determined volume fractions are correct.

EXAMPLE 15.7: CALCULATING THE STIFFNESS OF MMC

By using the data in Example 15.6, compute the stiffness (modulus of elasticity) of the MMC. The stiffness data are given as follows: $E_{Al} = 71$ GPa, $E_B = 370$ GPa, and $E_W = 410$ GPa.

Solution

$V_{Al} = 0.48,\ V_W = 0.006,\ V_B = 0.514,\ E_{MMC} = ?$

By using Equation 15.5,

$$E_{MMC} = E_{Al}V_{Al} + E_W V_W + E_B V_B = (71 \times 0.48) + (410 \times 0.006)$$
$$+ (370 \times 0.514) = 226.7\,\text{GPa}$$

The stiffness (modulus of elasticity) of the MMC $= 226.7$ GPa.

EXAMPLE 15.8: COMPUTING THE CURING RATIO FOR THE PULTRUSION PROCESS

A *pultrusion* composite processing is carried out by using an epoxy resin having the thermal activation energy of 50.6 kJ/mol. The curing temperature is 180°C, and the curing time duration is 20 min. The order of exothermic reaction for the resin is 1.89 and the pre-exponential constant is 6.6×10^4 /s. The universal gas constant, $R = 8.314$ J/mol.K. Compute the curing ratio for the pultrusion process.

Solution

$A_0 = 6.6 \times 10^4$/s, $E = 50.6$ kJ/mol $= 50600$ J/mol, $n = 1.89$,
$\qquad T = 180 + 273 = 453K,\ t = 1200\,s$

By using Equation 15.9,

$$k = A_0 \exp(-E/RT) = (6.6 \times 10^4)\,\exp[(-50600)/(8.314 \times 453)]$$
$$k = (6.6 \times 10^4)\,\exp(-13.43) = 6.6 \times 10^4 \times 1.46 \times 10^{-6} = 0.096$$

By using Equation 15.10,

$$\alpha = 1 - [(n-1)\,kt + 1]^{1/(1-n)} = 1 - \{[(1.89-1)\,(0.096 \times 1200)] + 1\}^{1/(1-1.89)}$$
$$= 1 - \{[(0.89)\,(115.2)] + 1\}^{1/(-0.89)} = 1 - (103.5)^{-1.12}$$
$$= 1 - 0.0055 = 0.9945$$

Hence, the curing ratio $= 0.9945$.

EXAMPLE 15.9: CALCULATING THE EFFECTIVE CURING RATE FOR THE PULTRUSION PROCESS

A polyurethane resin was used for the pultrusion process. The curing temperature was 210°C. Calculate the effective curing rate for the pultrusion process.

Solution

$R = 8.31\ \mathrm{J/mol \cdot K}$, $T = 210 + 273 = 483\ \mathrm{K}$

By using Equation 15.11,

$$k_{eff} = 0.51\ \exp\{(-57.7 \times 10^6)\,(1/R)\,[(1/T) - (1/373)]\}/\mathrm{min}$$

$$k_{eff} = 0.51\ \exp\{(-57.7 \times 10^6)\,(1/8.31)\,[(1/483) - (1/373)]\}$$

$$= 0.51\ \exp[(-57.7 \times 10^6)\,(0.12)\,(0.00207 - 0.00268)]$$

$$= 0.51\ \exp[(6.92)\,(0.00061)] = 0.51\ \exp(0.0042)$$

$$k_{eff} = 0.51 \times 1.0042 = 0.5121$$

Hence, the effective curing rate $= 0.5121/\mathrm{min}$.

EXAMPLE 15.10: CALCULATING VOLUME FRACTION OF VOIDS IN THE PRE-SILVER-INFILTRATED W COMPACT

A silver-tungsten particulate composite was manufactured by forming a porous tungsten powder compact followed by sintering and silver infiltration (see Figure 15.3). The pre-silver-infiltration density of the tungsten compact was measured to be 14.7 g/cm³. Calculate the volume fraction of voids, assuming that all voids are interconnected. The densities of commercially pure tungsten and pure silver are 19.3 g/cm³ and 10.5 g/cm³, respectively.

Solution

$\rho_w = 19.3 \text{ g/cm}^3, \rho_{Ag} = 10.5 \text{ g/cm}^3, \rho_{Compact} = 14.7 \text{ g/cm}^3, V_{voids} = ?$

Considering voids as a constituent, we can apply RoM (Equation 15.12) as follows:

$$\rho_{Compact} = V_{voids} \, \rho_{void} + V_W \rho_w$$

$$14.7 = (V_{voids} \times 0) + (V_W \times 19.3)$$

$$14.7 = 0 + 19.3 \, V_W$$

$$V_W = 0.76$$

By using Equation 15.13,

$$V_{voids} + V_W = 1$$

$$V_{voids} = 1 - V_W = 1 - 0.76 = 0.24$$

The volume fraction of voids $= V_{voids} = 0.24$.

EXAMPLE 15.11: CALCULATING THE POST-INFILTRATED WEIGHT % OF SILVER IN THE Ag-W COMPACT

By using the data in Example 15.10, calculate the post-infiltrated weight % of silver in the compact.

Solution

In the post-infiltrated compact, silver occupies all space that was previously occupied by voids in the compact, i.e., $V_{Ag} = V_{voids} = 0.24$.

Now, weight % Ag in the composite can be expressed as

Weight % Ag $= \{[(V_{Ag}) \, (\rho_{Ag})]/[(V_{Ag}) \, (\rho_{Ag}) + [(V_W) \, (\rho_w)]\} \times 100$

$= \{[(0.24 \times 10.5)]/[(0.24 \times 10.5) + (0.76 \times 19.3)]\} \times 100$

$= [(2.52)/(2.52 + 14.67)] \times 100 = [(2.52)/(17.2)] \times 100$

Weight % Ag $= 14.6$

Hence, the post-infiltrated weight % of silver in the compact $= 14.6\%$.

EXAMPLE 15.12: COMPUTING THE NUMBER OF ThO$_2$ PARTICLES IN EACH cm^3 OF A NANOCOMPOSITE

A TD-nickel nanocomposite is produced by adding 2.8 wt% thoria (ThO$_2$) to nickel. In the sintered compact, each nanosized ThO$_2$ particle has a diameter of 98 nm. Compute the number of ThO$_2$ particles in each cubic centimeter of the nanocomposite.

Solution

Wt% thoria $= 2.8$; weight of thoria in each gram of composite $= 2.8\% = 0.028$ g.

Wt% Ni $= 100 - 2.8 = 97.2$; weight of Ni in each gram of composite $= 0.972$ g.

Density of thoria $= \rho_{thoria} = 9.69$ g/cm^3.

Density of Ni $= \rho_{Ni} = 8.9$ g/cm^3.

Diameter of thoria nanoparticle $= 98$ nm $= 98 \times 10^{-9}$ m $= 98 \times 10^{-7}$ cm.

Thoria particle radius $= r = 0.5 \times 98 \times 10^{-7}$ cm $= 49 \times 10^{-7}$ cm.

$$V_{thoria} = (\text{wt. thoria}/\rho_{thoria}) = (0.028/9.69) = 0.00289 \text{ cm}^3.$$

$$V_{Ni} = (\text{wt. Ni}/\rho_{Ni}) = (0.972/8.9) = 0.1092 \text{ cm}^3.$$

Volume fraction of thoria $= f_{thoria} = (V_{thoria})/(V_{thoria} + V_{Ni})$.

$$f_{thoria} = (0.00289)/(0.00289 + 0.1092) = 0.00289/0.112 = 0.0258.$$

Volume occupied by thoria particles in each cm^3 of composite $= 0.0258$ cm^3.

Volume of a spherical thoria particle $= (4/3)\pi r^3 = (4/3)\pi(49 \times 10^{-7} \text{ cm})^3$.

Volume of a spherical thoria particle $= 4.18 \times 117649 \times 10^{-21}$ cm$^3 = 4.92 \times 10^{-16}$ cm^3.

Number of thoria particles in each cm^3 of the composite $= (0.0258 \text{ cm}^3)/(4.92 \times 10^{-16} \text{ cm}^3)$.

Number of thoria particles in each cm^3 of the composite $= 0.00524 \times 10^{16} = 5.24 \times 10^{13}$.

EXAMPLE 15.13: CALCULATING THE DENSITY OF CEMENTED CARBIDE COMPOSITE

A cemented carbide cutting tool is made of 73 wt% tungsten carbide, 16 wt% titanium carbide, 5 wt% tantalum carbide, and 6 wt% cobalt. Calculate the density of the composite.

Solution

$$V_1 = 73\% = 0.73, \ V_2 = 16\% = 0.16, \ V_3 = 5\% = 0.05, \ V_4 = 6\% = 0.06.$$

Density of WC $= \rho_1 = 15.8$ g/cm^3; Density of TiC $= \rho_2 = 4.9$ g/cm^3.

Density of TaC $= \rho_3 = 14.5$ g/cm^3; Density of Co $= \rho_4 = 8.9$ g/cm^3.

$$\text{Density of composite} = \rho_c = ?$$

By using Equation 15.12,

$$\rho_c = V_1\rho_1 + V_2\rho_2 + V_3\rho_3 + V_4\rho_4$$
$$= (0.73 \times 15.8) + (0.16 \times 4.9) + (0.05 \times 14.5) + (0.06 \times 8.9)$$

$$\rho_c = 11.53 + 0.78 + 0.72 + 0.53 = 13.56 \text{ g/cm}^3$$

The density of the composite $= 13.56$ g/cm^3.

Questions and Problems

15.1 Classify composites on the basis of (a) matrix materials and (b) reinforcement form.

15.2 Explain the technological importance of fiber orientation in FR composites.

15.3 What is a MMC? Explain the process of manufacturing a MMC.

15.4 (a) List the various thermoset composite manufacturing techniques. (b) Explain two methods of manufacturing a thermoset matrix composite.

15.5 What are particulate reinforced composites? Give four examples of the composites.

15.6 Explain the manufacturing process of the following particulate composites: (a) concrete and (b) silver-tungsten electrical composite.

15.7 What are dispersion strengthened nanocomposites? How are they manufactured?

P15.8 The elastic modulus of a glass fiber is 73 GPa, and its tensile strength is 2.8 GPa. A GFRP has been produced by reinforcing the glass fibers into the polymer matrix. The GFRP composite has 62% by volume of continuous glass fibers and a hardened epoxy resin having elastic modulus and tensile strength of 3.2 GPa and 64 MPa, respectively. Determine the elastic modulus of the GFRP composite under iso-strain condition.

P15.9 By using the data in 15.8 calculate the tensile strength of the GFRP composite.

P15.10 An MMC is made of an *Al* alloy matrix and continuous boron fibers. The boron fibers are produced using a 13 μm diameter tungsten-wire core that is coated with boron. The final diameter of the boron fiber is 110 μm. The volume fraction of the boron fiber is 53%. The boron fibers are oriented in the iso-strain condition in the *Al* alloy matrix. Calculate the volume fractions of (a) aluminum alloy matrix, (b) tungsten wire, and (c) boron.

P15.11 A new MMC is made of titanium aluminide and continuous silicon carbide (*SiC*) fibers oriented in the iso-strain condition. What must be volume fraction of *SiC* if the elastic modulus of the composite is 255 GPa? The stiffness data are given as follows: $E_{Ti3Al} = 145$ GPa and $E_{SiC} = 390$ GPa.

P15.12 A *pultrusion* composite processing is carried out by using an epoxy resin having the thermal activation energy of 51 kJ/mol. The curing temperature is 190°C and the curing time duration is 22 min. The order of exothermic reaction for the resin is 1.89 and the pre-exponential constant is 6.6×10^4/s. The universal gas constant, $R = 8.314$ J/mol.K. Compute the curing ratio for the pultrusion process.

P15.13 A polyurethane resin was used for the pultrusion process. The curing temperature was 200°C. Calculate the effective curing rate for the pultrusion process.

P15.14 A silver-tungsten particulate composite was manufactured by forming a porous tungsten powder compact followed by sintering and silver infiltration. The pre-silver-infiltration density of the tungsten compact was measured to be 14.7 g/cm^3. Calculate the volume fraction of voids, assuming that all voids are interconnected. The densities of commercially pure tungsten and pure silver are 19.3 g/cm^3 and 10.5 g/cm^3, respectively.

P15.15 A cemented carbide cutting tool is made of 77 wt% tungsten carbide, 14 wt% titanium carbide, 4 wt% tantalum carbide, and 5 wt% cobalt. Calculate the density of the composite.

References

Callister, W.D., Jr. 2007. *Materials Science and Engineering—An Introduction*. New York, NY: Wiley.

Chachad, Y.R., Roux, J.A., Vaughan, J.G., Arafat, E. 1995. Three-dimensional characterization of pultruded fiberglass-epoxy composite materials. *Journal of Reinforced Plastic and Composites*, 14, 495–512.

Hardis, R. 2012. Cure kinetics characterization and monitoring of an epoxy resin for thick composite structures. Master of Science Thesis, Iowa State University.

Huda, Z. (Ed.). 2017. Composite processing technology. In: *Foundations of Materials Science and Engineering*, Switzerland: Trans Tech Publications, 279–297.

Huda, Z., Edi, P. 2013. Materials selection in design of structures and engines of supersonic aircrafts: A review. *Materials and Design*, 46, 552–560.

Joseph, K., Thomas, S., Pavithran, C., Brahmakumar, M. 1993. Tensile properties of short sisal fiber reinforced polyethylene composites. *Journal of Applied Polymer Science*, 47(10), 1731–1739.

Mazumdar, S.K. 2002. *Composites Manufacturing: Materials, Product, and Process*. Boca Raton, FL: CRC Press.

Section II

Advanced Manufacturing

16

Nontraditional Machining Processes

16.1 Classification of Nontraditional Machining Processes

Nontraditional machining (NTM) are the processes that involve removal of excess material by various techniques that do not use sharp cutting tools as generally used in traditional machining processes. Additionally, no chips are produced during the NTM process. All NTM processes can be classified into four categories: (1) mechanical-energy machining (MEM), (2) thermal-energy machining (TEM), (3) chemical machining (ChM), and (4), electro-chemical machining (ECM) (Grzesik, 2016). These categories of NTM processes are discussed in the subsequent sections.

16.2 Mechanical-Energy Machining Processes

MEM processes involve the removal of unwanted material from the work-piece by an abrasive mechanical energy action. The MEM processes can be divided into five types: (1) abrasive jet machining (AJM), (2) ultrasonic machining (USM), (3) water jet machining (WJM), (4) abrasive water jet machining (AWJM), and (5) abrasive flow machining (AFM). These processes are discussed in the following sections.

16.2.1 Ultrasonic Machining

USM Process: USM is also called ultrasonic vibration machining. USM involves the application of a low-frequency electrical signal to a transducer, which converts the electrical energy into high-frequency mechanical vibration with a frequency in the range of 19–25 kHz (see Figure 16.1). This high-frequency mechanical energy is transmitted to a horn and tool assembly, which results in a unidirectional vibration of the tool at the ultrasonic frequency with a known amplitude in the range of 15–50 μm. A constant stream of abrasive slurry is also passed between the tool and workpiece. The vibrating tool, combined with the abrasive slurry mechanically erodes the material, thereby

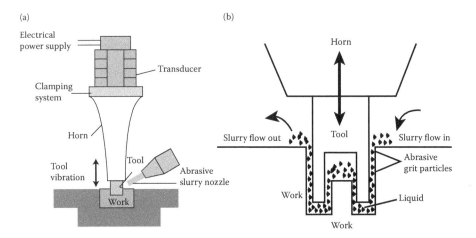

FIGURE 16.1
USM: (a) working principle and (b) abrasive grit motion.

leaving a precise reverse image of the tool shape. The USM process is capable of machining hard and brittle materials.

Mathematical Modeling: In USM, there are two principal mechanisms of material removal due to grit (abrasive particle) action: material removal due to grit throw and material removal due to grit hammering (Shaw, 1956).

The material removal rate (MRR) due to grit throw can be computed by

$$\text{MRR}_{th} = k_1 \, k_2 \, k_3 \left[v(h_{th})^3 / d_a \right]^{1/2} \tag{16.1}$$

where MRR_{th} is the material removal rate due to grit throw, mm^3/s; k_1, k_2, k_3 are the constants of proportionality; h_{th} is the depth of penetration due to grit throw, mm; v is the frequency of the tool oscillation, Hz; and d_a is the abrasive grit diameter, mm (see Example 16.2).

The depth of penetration of an abrasive particle (grit) can be calculated by

$$h_{th} = \pi \, a_t \, v \, d_a \, [\rho_a^3 / (6\sigma_w)]^{1/2} \tag{16.2}$$

where h_{th} is the depth of penetration due to grit throw, mm; a_t is two times the amplitude of tool oscillation, mm; ρ_a is the density of abrasive grit, g/mm^3; and σ_w is the mean stress acting on the work surface, MPa (see Example 16.1).

The material removal rate by grit hammering (MRR_h) can be determined by

$$\text{MRR}_h = k_1 \, k_2 \, k_3 \, [(h_h)^3 / d_a]^{1/2} v \tag{16.3}$$

where h_h is the depth of penetration due to grit hammering (see Example 16.4).

The depth of penetration due to grit hammering can be computed by

$$h_h \cong h_w = \{[(4\,F_f\,a_t\,d_a)/[(\pi\,k_2\,(j+1)\,\sigma_w]\}^{1/2} \tag{16.4}$$

where h_w is the depth of workpiece penetration due to grit hammering; F_f is the feed force on the tool, N; and j is the ratio of the mean stress on the work to the mean stress acting on the tool ($j = \dfrac{\sigma_w}{\sigma_t}$) (see Example 16.3).

Thus, the total volume of material removed per second can be determined by (El-Hofy, 2005)

$$\text{MRR} = \text{MRR}_{th} + \text{MRR}_h \tag{16.5}$$

where MRR is the total volume of material removed per second (i.e., the MRR), mm^3/s. The machining time for producing a hole by USM can be computed by

$$t_m = \frac{\text{Volume of hole}}{\text{MRR}} = [(\pi/4)D^2 t]/\text{MRR} \tag{16.6}$$

where t_m is the machining time, s; D is the hole diameter, mm; and t is the work thickness, mm. The significance of Equation 16.6 is illustrated in Example 16.5.

16.2.2 Water Jet Machining

WJM involves the use of a high-velocity water jet to cut and machine soft and nonmetallic materials (such as foam, plastic, etc.). In WJM, first water from the fluid supply is pumped to the intensifier using a hydraulic pump, which increases the pressure of the water to the required level (~300 MPa). Then the pressurized water is passed to an accumulator, where pressurized water is temporarily stored. The pressurized water then passes through a nozzle by use of control valve and flow regulator. At the nozzle, there is a tremendous increase in the kinetic energy of pressurized water. When the high-energy water jet strikes the workpiece, stresses are induced that result in the removal of material from the workpiece. The WJM process is capable of cutting materials in a sandwich up to 10 cm in thickness.

16.2.3 Abrasive Water Jet Machining

Principle: AWJM involves the use of a fine jet of very high-pressure water and abrasive slurry to more effectively cut the target material (including steels and other metals) by means of erosion (Gupta et al., 2014). In AWJM, the most commonly used abrasive is red garnet. The working principle of AWJM is similar to WJM except that an abrasive slurry is mixed with water in the former technique. The performance of AWJM depends on several factors, including the abrasive grain size. In order to achieve a higher MRR, a coarser abrasive with grit/mesh number in the range of 60–80 should be used. A finer abrasive with mesh number in the range of 100–150 is recommended to obtain a smoother surface finish (lower surface roughness, R_a).

Mathematical Modeling: The surface roughness of the material machined by AWJM depends on several factors, including abrasive grit size, stand-off distance, water jet pressure, work thickness, abrasive flow rate, and feed rate. Janković and coworkers have reported a useful mathematical model to estimate the surface roughness, as follows (Janković et al., 2012):

$$R_a = 0.2913\,[(s^{0.694} \cdot v^{0.648})/q^{0.212}] \tag{16.7}$$

where R_a is the surface roughness, µm; s is the work thickness, mm; v is the feed rate, mm/min; and q is the abrasive flow rate, g/min (see Example 16.6).

16.2.4 Abrasive Jet Machining

Principle: AJM involves the use of high-velocity abrasive particles that strike on the workpiece and remove the material by microcutting action as well as brittle fracture of the work material (see Figure 16.2). The performance of the AJM process strongly depends on the following process parameters: abrasive grain size, abrasive mass flow rate, nozzle tip distance, gas pressure, velocity of abrasive particles, and mixing ratio. The optimum ranges for these process parameters are presented in Table 16.1.

Mathematical Modeling: In AJM, the *mixing ratio* is the ratio of the volume flow rate of abrasive particles to the volume flow rate of air

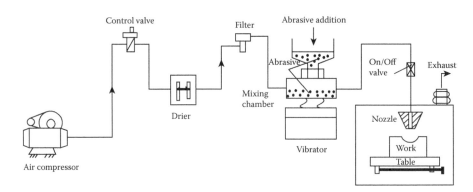

FIGURE 16.2
Abrasive jet machining.

TABLE 16.1

Optimum Process Parameters for Abrasive Jet Machining (AJM)

Process Parameter	Abrasive Grain Size	Abrasive Velocity	Nozzle Tip Distance	Nozzle ID
Value range	30–70 μm	150–300 m/s	0.75–2 mm	0.2–0.8 mm

(or gas). Numerically,

$$\text{Mixing ratio} = \frac{\text{Volume flow rate of abrasive particle}}{\text{Volume flow rate of carrier gas}} = V_a/V_g \qquad (16.8)$$

The mass ratio is defined as

$$\text{Mass ratio} = \frac{\text{Mass flow rate of abrasive particles}}{\text{Combined mass flow rate of carrier gas}} = M_a/(M_a + M_g)$$
$$(16.9)$$

In AJM, the MRR is greatly influenced by the following factors: (1) abrasive mass flow rate, (2) abrasive velocity, (3) density of abrasive grit, and (4) flow stress acting on the work material. Accordingly, MRR can be calculated by (Huda, 2017)

$$\text{MRR}_{\text{brittle}} = [(m_a v^{1.5})]/[(\rho_a)^{0.25}(\sigma_w)^{0.75}] \qquad (16.10)$$

where $\text{MRR}_{\text{brittle}}$ is the MRR of the brittle material being machined, mm^3/s; m_a is the mass flow rate of abrasives, kg/s; v is the velocity of abrasive grits, mm/s; ρ_a is the density of abrasive grit, kg/mm^3; and σ_w is the flow stress acting on the work material, MPa.

16.3 Thermal Energy Machining Processes

TEM involves the application of heat energy to a small portion of the work surface, thereby causing removal of material by fusion and/or vaporization. There are four types of TEM processes: (1) electric discharge machining (EDM), (2) electric discharge wire cutting (EDWC), (3) electron beam machining (EBM), and (4) laser beam machining (LBM). These TEM processes are discussed in the following sections.

16.3.1 Electric Discharge Machining

Principle: EDM involves metal removal by means of electric spark erosion. In EDM, a dielectric liquid is subjected to an electric voltage by using two electrodes resulting in material removal from the metallic workpiece due to a series of rapidly recurring current discharges between the two electrodes. The tool-electrode is made cathode (−), whereas work-electrode is anode (+). The EDM is a suitable process for manufacturing cutting tools and metallic molds and dies (e.g., molds for plastic injection molding, extrusion dies, wire-drawing dies, forging and heading dies, and the like).

    Mathematical Model: In EDM, the two electrodes (work and tool) may be considered as a capacitor due to the use of dielectric liquid. The energy required to create spark (electric discharge) is called the *spark energy*. The *spark energy* can be computed by

$$E_s = (CV^2)/2 \qquad (16.11)$$

where E_s is the spark energy, J; C is the capacitance of the charging capacitor, Farads; and V is the applied voltage, volts (see Example 16.9).

16.3.2 Electric Discharge Wire Cutting

EDWC involves the use of a continuously spooling conductive wire (cathode) and deionized water (dielectric) for EDM of a workpiece (anode). A power supply generates rapid electric pulses that result in a discharge between the work and the electrode (wire), which causes the melting or vaporization of a minute piece of metal resulting in material removal.

16.3.3 Electron Beam Machining

Principle: EBM employs a high-energy beam of electrons to drill small-diameter (0.1–0.3 mm) holes or cut small slots in both soft and hard metallic materials with a large depth-to-width ratio of penetration (Figure 16.3). First, the workpiece is placed in a vacuum chamber (pressure $= 10^{-4}$ torr). When the electron gun (with a tungsten filament) is connected to a high voltage (~150,000 V), the tungsten filament is heated up to a temperature above 2500°C. The presence of anode causes ejection of electrons, which are focused by the field formed by the grid cup and by a magnetic lens system. The high velocity focused electrons strike a workpiece over a small circular area, thereby causing a correspondingly rapid increase in the temperature of the workpiece, to well above its boiling point. This thermal effect results in material removal.

    Mathematical Modeling: In EBM, the electrical energy of the electron gun is first transferred to an electron beam, and finally, the kinetic energy of the

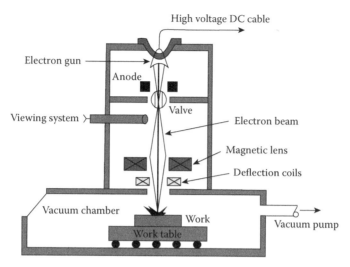

FIGURE 16.3
Electron beam machining.

electron beam is translated into heat energy of the work:

$$E = e \cdot V \qquad (16.12)$$

where E is the energy transferred to electron beam, J; e is the charge on an electron, Coulomb; and V is the potential difference between the filament and the anode, volts.

The electron starts from rest (from the electron gun) so the kinetic energy gained by an electron is

$$KE = \frac{1}{2} m \cdot v^2 \qquad (16.13)$$

where KE is the kinetic energy of electron beam, J; m is the mass of an electron, kg; and v is its speed, m/s. Each electron gains kinetic energy equal to the amount of energy transferred from the electron gun. By combining Equations 16.12 and 16.13, we obtain

$$\frac{1}{2} m \cdot v^2 = e \cdot V \qquad (16.14)$$

The significance of Equation 16.14 is illustrated in Example 16.10.
The cutting speed can be calculated by

$$v = (MRR)/A \qquad (16.15)$$

TABLE 16.2

Specific Power Consumption (C) Data for Various Metals

Material	Iron	Tungsten	Aluminum	Titanium
C, W/mm³/min	7	12	4	6

where v is the cutting speed, mm/min; MMR is the material removal rate, mm³/min; and A is the area of the slot or hole to be machined, mm².

The MRR can be found as

$$MRR = P/C \tag{16.16}$$

where P is the power of the electron beam, W; and C is the specific power, W/mm³/min. The C values for various metals are given in Table 16.2 (see Example 16.11).

The depth of penetration can be calculated by

$$d = 2.6 \times 10^{-17}(V^2/\rho) \tag{16.17}$$

where d is the depth of penetration, mm; V is the accelerating voltage, volts; and ρ is the density of the work material, kg/mm³ (see Example 16.13).

16.3.4 Laser Beam Machining

LBM involves the use of a high-energy coherent light beam to melt and vaporize particles on the surface of a metallic/nonmetallic workpiece. When a high-energy coherent-light laser beam is focused on a work surface, the thermal energy is absorbed, thereby transforming the work volume into a molten or vaporized state. The vaporized material is removed by flow of a high-pressure assisted gas jet.

16.4 Electrochemical Machining

Principle: ECM involves the removal of material by anodic metal dissolution. In ECM, the metallic workpiece is made the anode by connecting it to the positive (+) pole of a direct current (DC) generator. The tool is made the cathode (−) and is placed in front of the work area to be machined (see Figure 16.4). The gap between the two electrodes is filled with a conductive electrolytic solution (NaCl or NaNO$_3$), which handles the charge transfer in the working gap (Wilson, 1982).

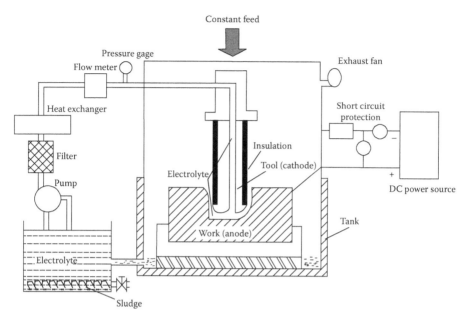

FIGURE 16.4
Electrochemical machining.

Mathematical Modeling: ECM is based on the principle of anodic metal dissolution; thus, Faraday's laws of electrolysis are applicable. The volume of metal dissolved per unit time (MRR) can be calculated by the following mathematical relationship:

$$\text{MRR} = \frac{A\,I}{\rho\,F\,Z} \qquad (16.18)$$

where MRR is the material removal rate, cm^3/s; A is the atomic mass of metal (work), g/mol; I is the current, ampere (A); ρ is the density of the metal (work), g/cm^3; Z is the valence of the metal; and F = Faraday's constant = 96,500 A-s (see Example 16.14).

The rate of dissolution can be computed by

$$dy/dt = (\text{MRR})/\text{tool area} \qquad (16.19)$$

where dy/dt is the rate of dissolution, mm/min; and MRR is the material removal rate, mm^3/min.

In ECM, the materials to be machined are generally alloys rather than pure metals. It is, therefore, more useful to develop an expression for the MRR of an alloy, as follows.

Assume there are k elements in an alloy.

The atomic masses of these elements can be represented as $A_1, A_2, \ldots\ldots,$ A_k with valence during electrochemical dissolution as $Z_1, Z_2, \ldots\ldots, Z_k$.

The mass percentages of the elements are $p_1, p_2, \ldots\ldots, p_k$ (in decimal fraction).

Accordingly, the MRR of the alloy can be calculated by

$$\text{MRR} = I/[F\rho\Sigma(p_iZ_i/A_i)] \tag{16.20}$$

where MRR is the material removal rate of the alloy, cm^3/s or mm^3/min; ρ is density of the alloy; and I is the current per unit valence of the lowest-valence metal in the alloy (see Example 16.15).

16.5 Chemical Machining

ChM involves the use of a strong acidic or alkaline chemical reagent for controlled chemical dissolution of the work material. In order to protect the areas from which the metal is not required to be removed, special coatings, called *maskants*, are used. ChM finds wise applications in the production of microcomponents of MEMS and semiconductor industry.

16.6 Problems and Solutions—Examples on Nontraditional Machining

EXAMPLE 16.1: CALCULATING THE DEPTH OF PENETRATION DUE TO GRIT THROW IN USM

A 5 mm diameter hole is produced in a tungsten carbide plate having thickness 1½ times the hole diameter by the USM process. The mean stress acting on the work surface is 6900 MPa. The mean abrasive grain diameter is 14 μm. The feed force on the tool is 3 N. The frequency of tool oscillation is 24 kHz, and the amplitude of tool vibration is 24 μm. The mean stress acting on the tool is 1300 MPa. The abrasive density is 4 g/cm³. The values of constants are $k_1 = 0.3$, $k_2 = 1.8 \text{ mm}^2$, $k_3 = 0.6$. Calculate the depth of penetration due to grit throw.

Solution

$a_t = 2 \times \text{amplitude} = 2 \times 24\,\mu\text{m} = 48 \times 10^{-6}\,\text{m} = 0.048\,\text{mm}$; $v = 24\,\text{kHz} = 24{,}000\,\text{Hz}$.

$d_a = 14\,\mu\text{m} = 14 \times 10^{-6}\,\text{m} = 0.014\,\text{mm}$; $\rho_a = 4\,\text{g/cm}^3 = 4 \times 10^{-3}\,\text{g/mm}^3$; $\sigma_w = 6900\,\text{MPa}$.

By using Equation 16.2,

$$h_{th} = \pi \, a_t \, v \, d_a \, [(\rho_a{}^3)/(6\sigma_w)]^{1/2}$$

$$= \pi \times 0.048 \times 24{,}000 \times 0.014[(4 \times 10^{-3})^3/(6 \times 6900)]^{1/2}$$

$$= 6.3 \times 10^{-5} \, \text{mm}$$

The depth of penetration due to grit throw $= h_{th} = 6.3 \times 10^{-5}$ mm.

EXAMPLE 16.2: COMPUTING THE MRR DUE TO GRIT THROW IN USM

By using the data in Example 16.1, compute the MRR due to grit throw.

Solution

$v = 24{,}000$ Hz, $h_{th} = 6.3 \times 10^{-5}$ mm, $(h_{th})^3 = 25 \times 10^{-14}$, $d_a = 0.014$ mm.
By using Equation 16.1,

$$\text{MRR}_{th} = k_1 \, k_2 \, k_3 \, [v(h_{th})^3/d_a]^{1/2}$$

$$= (0.3 \times 1.8 \times 0.6) \, [(24{,}000 \times 25 \times 10^{-14})/0.014]^{1/2}$$

$$= 0.324 \, (42{,}857{,}142.8 \times 10^{-14})^{1/2} = 2.12 \times 10^{-4} \, \text{mm}^3/\text{s}$$

The MRR due to grit throw $= \text{MRR}_{th} = 2.12 \times 10^{-4}$ mm^3/s.

EXAMPLE 16.3: CALCULATING THE DEPTH OF PENETRATION DUE TO GRIT HAMMERING IN USM

By using the data in Example 16.1, calculate the depth of penetration due to grit hammering in the USM process.

Solution

$F_f = 3$ N; $a_t = 0.048$ mm; $d_a = 0.014$ mm; $\sigma_w = 6900$ MPa; $\sigma_t = 1{,}300$ MPa, $k_2 = 1.8$ mm^2.

$$j = \frac{\sigma_w}{\sigma_t} = \frac{6900}{1300} = 5.3$$

By using Equation 16.4,

$$h_h \cong \{[(4 \, F_f \, a_t \, d_a)/[(\pi \, k_2 \, (j+1) \, \sigma_w]\}^{1/2}$$

$$= [(4 \times 3 \times 0.048 \times 0.014)/[(\pi \times 1.8 \times 6900(5.3+1)]^{1/2}$$

$$= [(0.0081)/(245817)]^{1/2} = 1.8 \times 10^{-4} \, \text{mm}$$

The depth of penetration due to grit hammering $= h_h \cong 1.8 \times 10^{-4}$ mm.

EXAMPLE 16.4: CALCULATING THE MRR BY GRIT HAMMERING IN USM

By using the data in Example 16.3, calculate the MRR by grit hammering in the USM process.

Solution

$k_1 = 0.3$, $k_2 = 1.8\,\text{mm}^2$, $k_3 = 0.6$; $v = 24{,}000\,\text{Hz}$; $d_a = 0.014\,\text{mm}$; $h_h = 1.8 \times 10^{-4}\,\text{mm}$.

By using Equation 16.3,

$$\text{MRR}_h = k_1\,k_2\,k_3\,[(h_h)^3/d_a]^{1/2}\,v = (0.3 \times 1.8 \times 0.6)\,[(1.8 \times 10^{-4})^3/\,0.014]^{1/2} \\ \times 24{,}000$$

$$\text{MRR}_h = 0.158\,\text{mm}^3/\text{s}$$

EXAMPLE 16.5: CALCULATING THE MACHINING TIME FOR A USM OPERATION

By using the data in Example 16.2 and Example 16.4, calculate the machining time for the USM operation.

Solution

$\text{MRR}_{th} = 2.12 \times 10^{-4}\,\text{mm}^3/\text{s}$; $\text{MRR}_h = 0.158\,\text{mm}^3/\text{s}$; $D = 5\,\text{mm}$.

Workpiece thickness $= t = 1.5\,D = 1.5 \times 5 = 7.5\,\text{mm}$.

By using Equation 16.5,

$$\text{MRR} = \text{MRR}_{th} + \text{MRR}_h = 2.12 \times 10^{-4} + 0.158 = 0.000212 + 0.158$$

$$\text{MRR} = 0.158212\,\text{mm}$$

By using Equation 16.6,

$$t_m = [(\pi/4)D^2\,t]/\text{MRR} = [(\pi/4)5^2 \times 7.5]/0.158212$$

$$t_m = 930.8\,\text{s} = 15.5\,\text{min}$$

The machining time for the USM operation $= 15.5\,\text{min}$.

EXAMPLE 16.6: CALCULATING THE SURFACE ROUGHNESS IN AWJM

A 6.8 mm thick metal plate is machined by AWJM by using a feed rate of 700 mm/min. The abrasive flow rate is 300 g/min. Estimate the surface roughness of the plate after machining.

Solution

$s = 6.8$ mm, $v = 700$ mm/min, $q = 300$ g/min.

By using Equation 16.7,

$$R_a = 0.2913 \, [(s^{0.694} \cdot v^{0.648})/q^{0.212}] = 0.2913 \, [(6.8)^{0.694} \, (700)^{0.648}/(300)^{0.212}]$$

$$= 0.2913 \, [(3.78 \times 69.76)/(3.35)] = 23 \, \mu m$$

The surface roughness of the plate after machining $= R_a = 23 \, \mu m$.

EXAMPLE 16.7: CALCULATING THE MASS RATIO WHEN THE MIXING RATIO IN AJM IS GIVEN

A mixing ratio of 0.23 is used in an AJM operation. The ratio of the density of the abrasive to the density of the gas is 18. Calculate the mass ratio.

Solution

Mixing ratio $= 0.23$, $\rho_a/\rho_g = 18$, $\rho_a = 18 \, \rho_g$; mass ratio $= ?$

By using Equation 16.8,

$$\text{Mixing ratio} = V_a/V_g$$

$$V_a/V_g = 0.23$$

$$V_a = 0.23 \, V_g$$

By using Equation 16.9,

$$\text{Mass ratio} = M_a/(M_a + M_g) = (\rho_a V_a)/(\rho_a V_a + \rho_g V_g)$$

$$= [(18 \, \rho_g) \, (0.23 \, V_g)]/[(18 \, \rho_g \cdot 0.23 \, V_g) + (\rho_g V_g)] = 0.805$$

Mass ratio $= 0.805$

EXAMPLE 16.8: COMPUTING THE MRR IN AJM

A flow stress of 3.7 GPa acts on a brittle material during its AJM. The abrasive mass flow rate is 1.8 g/min, and the velocity is 220 m/s. The density of the abrasive grit is 2.9 g/cm^3. Compute the MRR.

Solution

$m_a = 1.8$ g/min $= 1.8 \times 0.001$ kg/60 s $= 0.00003 = 3 \times 10^{-5}$ kg/s.

$v = 220$ m/s $= 220 \times 10^3$ mm/s $= 220{,}000$ mm/s.

$\rho_a = 2.9$ g/cm$^3 = 2.9 \times 0.001$ kg/(10 mm)$^3 = 2.9 \times 10^{-3}$ kg/10^3 mm$^3 = 2.9 \times 10^{-6}$ kg/mm^3.

$\sigma_w = 3.7$ GPa $= 3.7 \times 10^3$ MPa $= 3700$ MPa.

By using Equation 16.10,

$$\text{MRR}_{\text{brittle}} = [(m_a \, v^{1.5})]/[(\rho_a)^{0.25} \, (\sigma_w)^{0.75}] = [(3 \times 10^{-5}) \, (220{,}000)^{1.5}]/$$
$$[(0.0000029)^{0.25} \, (3700)^{0.75}]$$
$$= (309{,}567{,}440.1 \times 10^{-5})/(0.0412 \times 474.4)$$
$$= (3095.67)/(19.54) = 158.4 \text{ mm}^3/\text{s}$$

The MRR $= 158.4 \text{ mm}^3/\text{s}$.

EXAMPLE 16.9: CALCULATING THE SPARK ENERGY IN EDM

A metallic workpiece is machined by EDM by using an RC-type generator that applies the maximum charging voltage of 90 volts. The capacitance of the charging capacitor is 120 microfarad. Calculate the spark energy.

Solution

$C = 120 \, \mu\text{F} = 120 \times 10^{-6} \, \text{F} \; V = 90 \text{ volts } E_s = ?$

By using Equation 16.11,

$$E_s = \tfrac{1}{2} \, CV^2 = (0.5 \times 120 \times 10^{-6}) \, (90)^2 = 486{,}000 \times 10^{-6} \, \text{J} = 0.48 \, \text{J}$$

The spark energy $= 0.48 \, \text{J}$.

EXAMPLE 16.10: CALCULATING THE VELOCITY OF ELECTRONS IN EBM

An EBM operation is performed on a metallic work. The voltage between cathode and anode of an electron gun is 8×10^4 V. Calculate the velocity of electron in the electron beam.

Solution

$V = 8 \times 10^4$ volts; $m = 9.1 \times 10^{-31}$ kg; $e = 1.6 \times 10^{-19}$ coulomb.

By using Equation 16.14,

$$\tfrac{1}{2} \, mv^2 = eV$$

$$(0.5) \, (9.1 \times 10^{-31}) \, v^2 = 1.6 \times 10^{-19} \times 8 \times 10^4$$

$$4.55 \, v^2 = 12.8 \times 10^{-15}$$

$$v = 0.53 \times 10^{-7} = 5.3 \times 10^{-8} \, \text{m/s}$$

The velocity of electrons $= 5.3 \times 10^{-8} \, \text{m/s}$.

EXAMPLE 16.11: CALCULATING THE MRR IN EBM

A 230 µm wide slot is cut in a 0.8 mm thick titanium plate using a 1.1 kW electron beam during EBM. Calculate the MRR.

Solution

By reference to Table 16.2, $C = 6$; $P = 1.1 \times 1000 = 1100$ W.

By using Equation 16.16,

$$\text{MRR} = P/C = 1100/6 = 183.3 \, \text{mm}^3/\text{min}$$

The material removal rate $= \text{MRR} = 183.3 \, \text{mm}^3/\text{min}$.

EXAMPLE 16.12: CALCULATING THE CUTTING SPEED IN EBM

By using the data in Example 16.11, calculate the cutting speed in the EBM process.

Solution

Area of slot $= A = 0.8 \times 230 \times 10^{-3} \, \text{mm}^2 = 0.184 \, \text{mm}^2$ MRR $= 183.3$ mm^3/min.

By using Equation 16.5,

$$v = (\text{MRR})/A = 183.3/0.184 = 996.3 \, \text{mm/min}$$

The cutting speed $= 996.3 \, \text{mm/min}$.

EXAMPLE 16.13: CALCULATING THE DEPTH OF PENETRATION IN EBM

An electron beam with an accelerating voltage of 95 kV strikes on an aluminum plate during EBM. Calculate the depth of penetration.

Solution

$V = 95 \, \text{kV} = 95,000 \, \text{V}$ Density of aluminum $= \rho = 2.7 \, \text{g/cm}^3 = 2.7 \times 10^{-6}$ kg/mm^3.

By using Equation 16.17,

$$d = 2.6 \times 10^{-17} \, (V^2/\rho) = 2.6 \times 10^{-17} \, [(95,000)^2/(2.7 \times 10^{-6})]$$

$$= 8.7 \times 10^{-2} \, \text{mm} = 8.7 \times 10^{-5} \, \text{m} = 87 \times 10^{-6} \, \text{m} = 87 \, \mu\text{m}$$

The depth of penetration $= 87 \, \mu\text{m}$.

EXAMPLE 16.14: CALCULATING THE CURRENT REQUIREMENT FOR ECM WHEN MRR IS GIVEN

A MRR of 570 mm^3/min is required in electrochemical machining of nickel.

Calculate the current requirement for the ECM process.

Solution

MRR $= 570$ mm^3/min $= [570$ $(0.1$ cm$)^3]/60$ $s = 0.0095$ cm^3/s.

$A_{Ni} = 58.7$ g/mol, $Z_{Ni} = 2$, $F = 96,500$ coulomb, $\rho = 8.9$ g/cm^3.

By using Equation 16.18,

$$MRR = (A\ I)/(\rho\ F\ Z)$$

$$0.0095 = (58.7\ I)/(8.9 \times 96,500 \times 2)]$$

$$I = (0.0095 \times 8.9 \times 96,500 \times 2)/58.7 = 278 \text{ amperes}$$

The required current $= 278$ A.

EXAMPLE 16.15: CALCULATING MRR AND RATE OF DISSOLUTION OF A METAL IN ECM

In ECM of pure titanium, a 700 ampere current is passed. Calculate (a) the MRR and (b) the rate of dissolution if the tool area is 1000 mm^2.

Solution

$I = 700$ A, $A_{Ti} = 47.9$ g/mol, $\rho_{Ti} = 4.51$ g/cm^3, $Z_{Ti} = 3$, $F = 96,500$ A-s; Tool area $= 1000$ mm^2.

(a) By using Equation 16.18,

$$MRR = (A\ I)/(\rho\ F\ Z) = (47.9 \times 700)/(4.51 \times 96,500 \times 3) = 0.026 \text{ cm}^3/s$$

$$MRR = 0.026\ (10 \text{ mm})^3/ (1/60) \text{ min} = 1541 \text{ mm}^3/\text{min}$$

Metal removal rate $= MRR = 1541$ mm^3/min.

(b) By using Equation 16.19,

$$dy/dt = (MRR)/(\text{tool area}) = 1541/1000 = 1.54 \text{ mm-min}$$

Rate of dissolution $= 1.54$ mm-min.

EXAMPLE 16.16: CALCULATING MRR FOR STAINLESS STEEL IN ECM

Calculate the MRR in machining of 18–8 stainless steel by passing a current of 1700 amperes in ECM.

Solution

The 18–8 stainless steel contains 18% Cr and 8% Ni and balance iron (Fe).

$p_{Ni} = 8\% = 0.08$, $p_{Cr} = 18\% = 0.18$, $p_{Fe} = 74\% = 0.74$

By reference to the Periodic Table of Elements and the Physical Properties data, we obtain

$A_{Ni} = 58.71$ g/mol, $\rho_{Ni} = 8.9$ g/cm³, $Z_{Ni} = 2$; $A_{Cr} = 51.99$, $\rho_{Cr} = 7.19$ g/cm³, $Z_{Cr} = 2$;

$A_{Fe} = 55.85$ g/mol, $\rho_{Fe} = 7.86$ g/cm³, $Z_{Fe} = 2$

$I = $ (current)/(valence of the lowest valence metal) $= 1700/2 = 850$ ampere

The density of the stainless steel can be determined as follows:

$$\rho_{alloy} = 1/\Sigma \ (p_i/\rho_i)$$

$$\rho_{alloy} = 1/[(0.08/8.9) + (0.18/7.19) + (0.74/7.86)] = 7.81 \text{ g/cm}^3$$

By using Equation 16.20,

$$MRR = I/[F \ \rho \ \Sigma \ (p_i \ Z_i/A_i)]$$

$$MRR = 850/\{(96{,}500 \times 7.81) \ [(0.08 \times 2/58.7) + (0.18 \times 2/52) + (0.74 \times 2/55.8)]\}$$

$$MRR = 0.0312 \text{ cm}^3/\text{s} = 0.0312 \times 1000 \times 60 = 1873.3 \text{ mm}^3/\text{min}$$

The MRR $= 1873.3$ mm³/min.

EXAMPLE 16.17: CALCULATING MRR FOR A SUPERALLOY IN ECM

A superalloy has composition: $Ni = 72$ wt%, $Cr = 17$ wt%, $Fe = 5$ wt%, $Al = 1$ wt%, and balance is titanium (Ti). The superalloy is machined by ECM by using a current of 1800 A through the cell. Calculate the MRR, assuming the current in the lowest valence metal.

Solution

$p_{Ni} = 72\% = 0.72$, $p_{Cr} = 17\% = 0.17$, $p_{Fe} = 0.05$, $p_{Al} = 0.01$, $p_{Ti} = 0.05$.

By reference to the Periodic Table of Elements and the Physical Properties data, we obtain

$A_{Ni} = 58.71$ g/mol, $\rho_{Ni} = 8.9$ g/cm³, $Z_{Ni} = 2$; $A_{Cr} = 51.99$, $\rho_{Cr} = 7.19$ g/cm³, $Z_{Cr} = 2$;

$A_{Fe} = 55.85$ g/mol, $\rho_{Fe} = 7.86$ g/cm³, $Z_{Fe} = 2$; $A_{Al} = 27$ g/mol, $\rho_{Al} = 2.7$ g/cm³, $Z_{Al} = 3$;

$A_{Ti} = 47.9$ g/mol, $\rho_{Ti} = 4.51$ g/cm³, $Z_{Ti} = 3$.

$$I = \text{(current)/(valence of the lowest valence metal)}$$
$$= 1800/Z_{lowest} = 1800/2 = 900 \text{ A}$$

The density of the superalloy can be determined as follows:

$$\rho_{alloy} = 1/\Sigma \, (p_i/\rho_i)$$

$$\rho_{alloy} = 1/[(0.72/8.9) + (0.17/7.19) + (0.05/7.86) + (0.01/2.7) + (0.05/4.51)] = 7.95 \, \text{g/cm}^3$$

By using Equation 16.20,

$$MRR = I/[F \, \rho \, \Sigma \, (p_i \, Z_i/A_i)]$$

$$MRR = 900/[(96{,}500 \times 7.95) \, [(0.72 \times 2/58.7) + (0.17 \times 2/52) + (0.05 \times 2/55.8) + (0.01 \times 3/27) + (0.05 \times 3/48)]$$

$$= 900/[(767{,}783.4) \, (0.0245 + 0.0065 + 0.0018 + 0.0011 + 0.0031)]$$

$$MRR = 0.0357 \, \text{cm}^3/\text{s} = 0.0357 \times 1000 \times 60 = 2144 \, \text{mm}^3/\text{min}$$

The metal removal rate = MRR = $2144 \, \text{mm}^3/\text{min}$.

Questions and Problems

16.1 Classify the various NTM processes with the aid of a classification chart.

16.2 You are given the following acronyms: ECM, USM, AWJM, ChM, EDM, AJM, WJM. (a) Give the meanings of each of the acronyms. (b) Arrange the acronyms under the four categories of NTM.

16.3 Draw a labeled sketch of USM.

16.4 Both EDM and ECM involve the use of electrolyte. What is the difference in the nature of the electrolyte used in the two processes?

16.5 Indicate which NTM process is the best for machining of each of the following materials: (a) high-strength alloy, (b) foam, (c) engineered ceramics, (d) ductile steel, (e) microcomponents, (f) drilling nanosize hole, and (g) brittle material with high MRR.

16.6 Explain AWJM with the aid of a diagram.

P16.7 The mean stress acting on the work surface, during an USM operation, is 6800 MPa. The mean abrasive grain diameter is 15 μm. The feed force on the tool is 2.9 N. The frequency of tool oscillation is 23 kHz, and the amplitude of tool vibration is 22 μm. The mean stress acting on the tool is 1100 MPa. The abrasive density is 3.8 g/cm³. The values of constants are $k_1 = 0.3$, $k_2 = 1.8 \, \text{mm}^2$, $k_3 = 0.6$. Calculate the depth of penetration due to grit throw.

P16.8 By using the data in P16.7, calculate the depth of penetration due to grit hammering.

P16.9 By using the data in P16.7, calculate the MRR by grit hammering, if the depth of penetration due to hammering is 1.7×10^{-4} mm.

P16.10 A flow stress of 3.5 GPa acts on a brittle material during AJM. The abrasive mass flow rate is 1.7 g/min, and the velocity is 190 m/s. The density of the abrasive grit is 3.1 g/cm^3. Compute the MRR.

P16.11 A 4 mm diameter hole is drilled in a workpiece in an USM operation; the work thickness is 1.5 times the hole diameter. The MRR due to grit throw is 0.0002 mm^3/s, and the MRR due to grit hammering is 0.16 mm^3/s. Compute the machining time.

P16.12 An electron beam with an accelerating voltage of 98 kV strikes on an aluminum plate during EBM. Calculate the depth of penetration.

P16.13 A 1.2 kW electron beam is used to machine a slot in a tungsten plate during EBM. Calculate the MRR.

P16.14 An 18–2 stainless steel is machined by passing a current of 2000 A in ECM. Compute (a) MRR and (b) rate of dissolution if the tool area is 1200 mm^2.

References

El-Hofy, H.A-G. 2005. *Advanced Machining Processes: Nontraditional and Hybrid Machining Processes*. New York, NY: McGraw-Hill.

Grzesik, W. 2016. *Advanced Machining Processes of Metallic Materials*, 2nd Edition. New York, NY: Elsevier Science.

Gupta, V., Pandey, P.M., Garg, M.P., Khanna, R., Batra, N.K. 2014. Minimization of *kerf* taper angle and kerf width using Taguchi's Method in abrasive water jet machining of marble. *Procedia Materials Science*, 6, 140–149.

Huda, Z. 2017. Advanced/Non-traditional Machining. In: (Ed. Z. Huda). *Materials Processing in Engineering Manufacture*. Zurich, Switzerland: Trans Tech, pp. 301–321.

Janković, P., Igić, T., Nikodijević, D. 2012. Process parameter effect on material removal mechanism and cut quality of abrasive water jet machining. *Theoretical and Applied Mechanics*, Series: Special Issue—*Address to Mechanics*, 40(S1), 277–291, Belgrade.

Shaw, M.C. 1956. Ultrasonic grinding. *Microtechnic*, 10(6), 257–265.

Wilson, J.F. 1982. *Practice and Theory of Electrochemical Machining*. Malabar, FL: R.E. Krieger.

17

Computer-Aided Manufacturing (CAD/CAM)

17.1 Role of Computer Numerical Control Machines in Computer-Aided Manufacturing

Computer-aided manufacturing (CAM) activities include the use of computer applications to develop a manufacturing plan for tooling design, computer-aided design (CAD) model preparation, numerical control (NC) programming, coordinate measuring machine (CMM) inspection programming, machine tool simulation, or postprocessing; the plan is then executed in a production environment.

Computer numerical controlled (CNC) machines play a pivotal role in CAM. CNC machines are electromechanical devices that manipulate machine tools using computer programming inputs that are fed into the machine control unit (MCU) of the machine (see Sections 17.3 and 17.4). A CNC machine interprets the design as instructions in CAD for machining a part made of any material. Commonly used work materials in a CNC machine include metals as well as wood, foam, fiberglass, and plastics. Some of the common machine tools that can run on the CNC are lathe, milling machine, drilling machine, and the like (Overby, 2010). Figure 17.1 shows a TMV-510CII CNC machine that is categorized as a *CNC Machining Center*. It can perform multiple machining operations including high-speed profile milling, drilling, and tapping. It is equipped with a C-type automatic tool changer (ATC) with a capacity of 14 tools having tool changing time of 2.8 seconds; each tool is assigned a specific serial number, which can be used to call that particular tool within the program (Qureshi, 2017).

17.2 Computer-Aided Design

17.2.1 What Is CAD?

CAD involves the use of a computer for creating computer models defined by geometrical parameters; these models typically appear on a computer monitor

FIGURE 17.1
A TMV-510CII CNC machine. (Courtesy of DHA Suffa University, Pakistan.)

as a three-dimensional (3D) representation of a part or a system of parts. A CAD software application enables the designer to analyze the computer model, modify it, and optimize the design so as to create a detailed production drawing. CAD systems enable designers to view products under a wide variety of representations (both 2D and 3D) and to test these products by simulating in a real-world environment.

17.2.2 Functions of CAD Software

A CAD software application should be able to perform the following functions: graphic shape generation, transformations, windowing, segmenting, and user input.

Graphic Shapes Generation: Graphic shapes are points, lines, arcs, circles, and other shapes that make up logos, illustrations, and countless other elements in all types of designs (see Figure 17.2). A CAD software package stores the information of various graphical shapes as well as alphanumeric elements. The graphic designer can generate different graphical shapes by specifying the dimensions and location of each individual shape, thereby creating a geometrical model by a combination (addition, subtraction, and grouping) of these shapes.

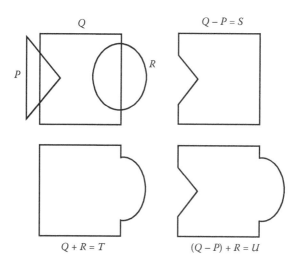

FIGURE 17.2
Graphic shapes generation in CAD.

Transformations: Transformations refer to the editing operations that are performed by the user to alter the shape, size, and/or orientation of the geometrical shapes. In general, three types of transformation are used: translation, rotation, and scaling (Fahad and Huda, 2017). *Translation* refers to moving the shape from one location to another. A line segment, X, is typically represented by the Cartesian coordinates (x, y) of its end points; which can be expressed in the matrix form as follows:

$$X = [x \quad y] \tag{17.1}$$

Translation is done by adding the desired values to the coordinates of the end point of the line:

$$X_t = X + T \tag{17.2}$$

where X is the original line matrix; T is the applied translation matrix $= \begin{bmatrix} a & b \\ c & d \end{bmatrix}$; and X_t is the translated line matrix $= [x_t \quad y_t]$. The significance of Equations 17.1 and 17.2 in the translation transformation function of CAD is illustrated in Example 17.1.

Another transformation function of CAD is rotation about the origin by an angle, θ. *Rotation* is performed by multiplying the original matrix with the rotational matrix as follows:

$$X_r = XR \tag{17.3}$$

where R is the applied rotation matrix $= \begin{bmatrix} \cos\theta & \sin\theta \\ -\sin\theta & \cos\theta \end{bmatrix}$; and X_r is the rotated line matrix. By substituting the values of X and R in Equation 17.3, we obtain

$$X_r = XR = \begin{bmatrix} x & y \end{bmatrix}\begin{bmatrix} \cos\theta & \sin\theta \\ -\sin\theta & \cos\theta \end{bmatrix} = \begin{bmatrix} x\cos\theta - y\sin\theta & x\sin\theta + y\cos\theta \end{bmatrix}$$

$$(17.4)$$

The significance of Equations 17.3 and 17.4 is illustrated in Example 17.2.

Scaling refers to increasing or reducing the shape size by a number. *Scaling* is performed by multiplying the original matrix with the scaling matrix, as follows:

$$\text{Scaling matrix} = S = \begin{bmatrix} x_s & 0 \\ 0 & y_s \end{bmatrix} \tag{17.5}$$

17.3 CNC System

Working Principles: A CNC system typically comprises the following main elements: (1) part program, (2) machine control unit (MCU), and (3) processing equipment (see Figure 17.3). A *part program* is a set of coded instructions to be followed by the processing equipment to produce a part. The part program is generally written by using a CAM software package (e.g., *NX CAM, CAM EXPRESS*, etc.). The MCU is the computer hardware that stores and executes the part program by issuing pulses and by implementing the required motions in the processing unit. The MCU issues a set of pulses that act as speed and feed axis motion signals to the amplifier circuits for driving the axis

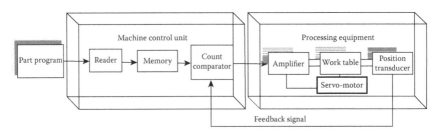

FIGURE 17.3
A CNC system—a closed-loop CNC system.

mechanisms. It also implements auxiliary control functions (e.g., coolant or spindle ON/OFF) and tool change (see the 14 tools turret facility in the CNC machine: Figure 17.1). The *processing equipment* (e.g., automatic lathe, automatic milling machine) performs the sequence of operations by using a drive system to transform the workpiece into a completed part. A drive system consists of amplifier circuits, stepping motors or servomotors, and ball screws. In an open-loop system, steeping motors are used, whereas the closed-loop system involves the use of servomotors (see Figure 17.3).

Open-Loop and Closed-Loop Systems: An open-loop CNC system does not involve feedback signal for error correction; hence, it operates without verifying that the desired position of the worktable has been achieved. A closed-loop CNC system operates by verifying that the desired position of the worktable has been achieved; here axis position feedback signals are issued by the *position transducer* to generate position feedback signals for error correction to a *comparator*, which compares the actual results with the desired results and issues corrective signals to the worktable via a digital-to-analog converter (DAC), amplifier, and servomotor (see Figure 17.3).

Mathematical Modeling: The MCU unit issues *pulses* to implement the required motion in the processing equipment (motor). *Each pulse* causes the stepping motor (or servomotor) to rotate a *fraction of one revolution*, called the *step angle*. The step angle can be computed by

$$\alpha = 360/n_s \qquad (17.6)$$

where $\alpha =$ the step angle, deg/step (open loop) $=$ angle between slots, deg/pulse (closed loop); $n_s =$ number of step angles for the motor (open-loop system) $=$ number of slots in the disk, slots/rev (closed loop system) $=$ number of pulses issued/rev of the ball screw (closed loop system).

The gear ratio is defined as the ratio of the rotational speed of the motor (N_m) to the rotational speed of the ball screw (N_b). Mathematically, the gear ratio can be expressed by

$$r_g = N_m/N_b = A_m/A_b \qquad (17.7)$$

where A_m is the angle of motor-shaft rotation, deg.; and A_b is the angle of ball-screw rotation, deg.

Modeling for Open-Loop System: The angle of motor-shaft rotation (A_m) can be computed by

$$A_m = n_p\alpha \qquad (17.8)$$

where α is the step angle, deg./step; and n_p is the number of pulses received by the motor.

The distance moved by the worktable in the x-direction can be calculated by

$$x = (p \cdot A_b)/360 \tag{17.9}$$

where x is the distance moved by the worktable in the x-direction, mm; p is the pitch of the ball screw, mm/rev; and A_b is the angle of ball-screw rotation, deg.

The number of pulses received by the motor (n_p) can be related to the gear ratio (r_g) as follows:

$$n_p = (360 \cdot x \cdot r_g)/(p \cdot \alpha) \tag{17.10}$$

The table travel speed or the feed rate (f_r, mm/min) can be calculated by

$$f_r = N_b \cdot p \tag{17.11}$$

where N_b is the rotational speed of the ball screw; and p is the pitch of the ball screw, mm/rev.

The pulse train frequency (f_p, Hz) can be related to the gear ratio (r_g) by

$$f_p = (f_r \cdot n_s \cdot r_g)/(60 \cdot p) \tag{17.12}$$

where n_s is the number of step angles of the stepper motor.

Modeling for Closed-Loop System: The distance moved by the worktable in the x-direction (x, mm) can be calculated by

$$x = (p \cdot n_p)/n_s = (p \cdot A_b)/360 \tag{17.13}$$

where n_p is the number of pulses sensed by count comparator; n_s is the number of pulses per revolution of the ball screw, issued by the MCU; and A_b is the angle of ball screw rotation, deg.

The rotational speed of the ball screw can be calculated by using Equation 17.11.

The number of pulses sensed by comparator (n_p) is related to the angle between slots (α, deg.) by

$$n_p = A_b/\alpha = (A_b \cdot n_s)/360 \tag{17.14}$$

The pulse train frequency (f_p, Hz) can be calculated by

$$f_p = (f_r \cdot n_s)/(60\,p) \tag{17.15}$$

where f_r is the feed rate or table travel speed, mm/min.

17.4 Part Programming in CAM

17.4.1 Code Words in Part Programming

A part program is the list of instructions to the CNC machine tool to perform operations for achieving the desired final geometry of the part (Smid, 2007). The commonly used *codes* have been standardized as ISO 6983. In a part program, a *block* refers to a combination of different words to indicate multiple control instructions (Manton and Weidinger, 2010). In order to prepare a *block* using a combination of words, the following sequence of codes is generally employed: N-word (for the sequence of block), G-word (for general instructions related to movements of tool/workpiece), F-word (for desired feed rate), S-word (for controlling spindle speed), T-word (for selecting desired tool), M-word (for different machine related functions; e.g., M07 for coolant ON), X-, Y-, Z- indicate linear axes; and A-, B-, C- indicate rotational axes.

17.4.2 G-words in Part Programming

G-words refer to common preparatory functions related to the movement of tool/work. Some commonly used G-words and their meanings are listed as follows:

G00—Used for positioning of the feed drives along a random path at a rapid traverse rate.

G01—Used for linear interpolation. For example, if the cutter head is currently located at $X = 3$ and $Y = 3$; and we want the tool to be moved to $X = 5$, $Y = 8$ with a feed rate of 120 mm/min, the syntax (using absolute positioning) would be G01 X5 Y8 F120 (see Example 17.10).

G02, G03—Used circular interpolation CW (G02) or CCW (G03). For example, if the cutter head is currently located at $X = 3$ and $Y = 3$ and we want it to move in a clockwise circular direction defined by $X = 7$ and $Y = 7$ with radius $= 10$ units, the syntax (using absolute positioning) would be G02 X7 Y7 R10 F120 (see Example 17.11).

G17, G19—Used to select planes (XY, XZ, and YZ, respectively) for circular interpolation, cutter compensation, and other similar functions as required.

G40, G41, G42—Used for incorporating the compensation of cutter diameter into the locations of the part program. G41 is used for compensating the cutter to the left, G42 is used for compensation on the right side of the part surface, and G40 is used to cancel the cutter compensation. For example, to compensate for a 12 mm diameter tool to the right side of the work while moving the tool to a position defined by $X = 20$ and $Y = 18$ would be written as follows: G42 G01 X20 Y18 D12 F120.

17.5 Problems and Solutions—Examples in CAD/CAM

EXAMPLE 17.1: PERFORMING THE TRANSLATION OPERATION AS A TRANSFORMATION FUNCTION OF CAD

A line, defined by coordinates of end points $(2, 5)$ and $(1, 3)$ and matrix X, is required to be translated by two units in x and three units in y directions. Perform the indicated translation operation.

Solution

By using Equation 17.1, the line is represented by matrix X, as follows:

$$X = \begin{bmatrix} 2 & 5 \\ 1 & 3 \end{bmatrix}$$

In order to perform translation, we use Equation 17.2,

$$X_t = X + T = \begin{bmatrix} 2 & 5 \\ 1 & 3 \end{bmatrix} + \begin{bmatrix} 2 & 3 \\ 2 & 3 \end{bmatrix} = \begin{bmatrix} 2+2 & 5+3 \\ 1+2 & 3+3 \end{bmatrix} = \begin{bmatrix} 4 & 8 \\ 3 & 6 \end{bmatrix}$$

EXAMPLE 17.2: PERFORMING THE ROTATION OPERATION AS A TRANSFORMATION FUNCTION OF CAD

The line, specified in Example 17.1, is required to be rotated about the origin by an angle of $20°$.
  Perform the indicated operation.

Solution

By using Equation 17.3,

$$X_r = XR = \begin{bmatrix} 2 & 5 \\ 1 & 3 \end{bmatrix} \begin{bmatrix} \cos 20 & \sin 20 \\ -\sin 20 & \cos 20 \end{bmatrix} = \begin{bmatrix} 2 & 5 \\ 1 & 3 \end{bmatrix} \begin{bmatrix} 0.94 & 0.34 \\ -0.34 & 0.94 \end{bmatrix}$$

$$X_r = \begin{bmatrix} (2 \times 0.94) - (5 \times 0.34) & (2 \times 0.34) + (5 \times 0.94) \\ (1 \times 0.94) - (3 \times 0.34) & (1 \times 0.34) + (3 \times 0.94) \end{bmatrix} = \begin{bmatrix} 0.18 & 5.38 \\ -0.08 & 3.16 \end{bmatrix}$$

EXAMPLE 17.3: PERFORMING SCALING OPERATION AS A TRANSFORMATION FUNCTION OF CAD

The line, specified in Example 17.1, is required to be scaled by three times in both x and y directions. Perform the indicated operation.

Solution

In order to increase (scale) the size by three times, we use Equation 17.5 by taking $x_s = 3$ and $y_s = 3$:

$$X_s = XS = \begin{bmatrix} 2 & 5 \\ 1 & 3 \end{bmatrix} \begin{bmatrix} x_s & 0 \\ 0 & y_s \end{bmatrix} = \begin{bmatrix} 2 & 5 \\ 1 & 3 \end{bmatrix} \begin{bmatrix} 3 & 0 \\ 0 & 3 \end{bmatrix} = \begin{bmatrix} 6 & 15 \\ 3 & 9 \end{bmatrix}$$

EXAMPLE 17.4: CALCULATING THE NUMBER OF PULSES REQUIRED IN AN OPEN-LOOP CNC SYSTEM

The worktable of an open-loop CNC system is driven by a ball screw that has 4 mm pitch. The table is connected to the output shaft of a stepper motor (having 45 step angles) through a gearbox whose gear ratio is 4:1. It is required to move the worktable a distance of 180 mm from its present position at table travel speed of 230 mm/min. Calculate the number of pulses required.

Solution

$p = 4$ mm, $n_s = 45$, $r_g = 4$, $x = 180$ mm, $n_p = ?$

By using Equation 17.6,

$$\alpha = 360/n_s = 360/45 = 8°$$

By using Equation 17.10,

$$n_p = (360 \cdot x \cdot r_g)/p \cdot \alpha = (360 \times 180 \times 4)/(4 \times 8) = 259{,}200/32 = 8100$$

The required number of pulses $= n_p = 8100$.

EXAMPLE 17.5: CALCULATING THE MOTOR SPEED TO MOVE THE TABLE IN OPEN-LOOP CNC SYSTEM

By using the data in Example 17.4, calculate the required motor speed.

Solution

$r_g = 4$, $n_p = 8100$, $f_r = 230$ mm/min, $p = 4$ mm, $N_m = ?$

By rearranging the terms in Equation 17.11,

$$N_b = f_r/p = 230/4 = 57.5 \text{ rpm}$$

By rearranging the terms in Equation 17.7,

$$N_m = r_g \cdot N_b = 4 \times 57.5 = 230 \text{ rpm}$$

EXAMPLE 17.6: CALCULATING THE PULSE TRAIN FREQUENCY IN AN OPEN-LOOP CNC SYSTEM

By using the data in Example 17.4, calculate the pulse train frequency corresponding to the desired table speed.

Solution

$r_g = 4$, $n_p = 8100$, $f_r = 230$ mm/min, $p = 4$ mm, $n_s = 45$.

By using Equation 17.12,

$$f_p = (f_r \cdot n_s \cdot r_g)/(60 \cdot p) = (230 \times 45 \times 4)/(60 \times 4) = 41{,}400/240 = 172.5\,\text{Hz}$$

The pulse train frequency $= 172.5\,\text{Hz}$.

EXAMPLE 17.7: CALCULATING THE NUMBER OF PULSES SENSED BY COMPARATOR IN CLOSED-LOOP SYSTEM

The ball screw of a closed-loop CNC system has a pitch of 4.5 mm and is coupled to a servomotor with a gear ratio of 4:1. The MCU of the CNC machine issues 90 pulses per revolution of the ball screw. The worktable is programmed to move a distance of 80 mm at a speed of 430 mm/min. Calculate the number of pulses sensed by the count comparator to verify the exact 80 mm movement of the worktable (see Figure 17.3).

Solution

$p = 4.5\,\text{mm}$, $r_g = 4$, $n_s = 90$ pulses/rev., $x = 8\,\text{mm}$, $f_r = 430\,\text{mm/min}$, $n_p = ?$
By rearranging the terms in Equation 17.13,

$$n_p = (x \cdot n_s)/p = (80 \times 90)/4.5 = 1600 \text{ pulses}$$

The number of pulses sensed by the count comparator $= n_p = 1600$ pulses.

EXAMPLE 17.8: CALCULATING THE PULSE TRAIN FREQUENCY IN CLOSED-LOOP CNC SYSTEM

By using the data in Example 17.7, calculate the pulse train frequency.

Solution

$p = 4.5\,\text{mm}$, $r_g = 4$, $n_s = 90$ pulses/rev., $x = 80\,\text{mm}$, $f_r = 430\,\text{mm/min}$, $f_p = ?$
By using Equation 16.15,

$$f_p = (f_r \cdot n_s)/(60\,p) = (430 \times 90)/(60 \times 4.5) = 143.3 \text{ cycles per second}$$

The pulse train frequency $= f_p = 143.3$ Hertz.

EXAMPLE 17.9: CALCULATING THE SERVOMOTOR ROTATIONAL SPEED IN CLOSED-LOOP CNC SYSTEM

By using the data in Example 17.7, calculate the servomotor rotational speed that corresponds to the specified table travel speed.

Solution

$p = 4.5\,\text{mm}$, $r_g = 4$, $n_s = 90$ pulses/rev., $x = 80\,\text{mm}$, $f_r = 430\,\text{mm/min}$, $N_m = ?$

By rearranging the terms in Equation 17.11,

$$N_b = f_r/p = 430/4.5 = 95.5 \, \text{rpm}$$

By rearranging the terms in Equation 17.7,

$$N_m = r_g \cdot N_b = 4 \times 95.5 = 382.2 \, \text{rpm}$$

EXAMPLE 17.10: WRITING SYNTAX FOR THE PART PROGRAM INVOLVING VARIOUS INTERPOLATIONS FOR CAM

The cutter head of the processing equipment of a CNC machine is currently located at $X = 2$ and $Y = 2$. Write the syntax (using absolute positioning) for the part program if it is desired to move the tool at a feed rate of 140 mm/min for the following cases: (a) tool straight line motion to $X = 6$ and $Y = 9$; (b) tool counterclockwise motion defined by $X = 6$, $Y = 6$, with radius $= 13$; and (c) to compensate for a 15 mm diameter tool to the left side of the workpiece while moving the tool to a position defined by $X = 17$ and $Y = 14$.

Solution

In order to write the syntax for the part program, we need to refer to Section 17.4.2.

(a) This motion involves linear interpolation. The syntax would be: G01 X6 Y9 F140.
(b) This motion involves circular interpolation. The syntax would be: G03 X6 Y6 R13 F140.
(c) This motion involves a combination of linear interpolation and compensation of cutter diameter. Hence, the syntax of the part program is as follows: G41 G01 X17 Y14 D15 F110.

EXAMPLE 17.11: WRITING A BLOCK OF A PART PROGRAM IN CAM

The cutter head (Tool 2) of a CNC machine's processing equipment is located at $X = 10$ and $Y = 3$, $Z = 2$. It is required to move the tool to $X = 20$, $Y = 7$, $Z = 3.8$ at a feed rate of 120 mm/min while coolant is ON. The spindle rotational speed is required to be 280 rev/min. Write the *block* (with sequence number 3) for the CNC part program.

Solution

This problem involves linear interpolation (see Section 17.4.2). Hence, the *block* for the part program would be as follows: N30 G01 X20 Y7 Z3.8 T02 S280 F120 M07.

EXAMPLE 17.12: DETERMINING COORDINATES OF MACHINING POINTS FOR PART PROGRAMMING

Figure E17.12 shows a billet with dimensions (300 mm × 300 mm × 25 mm). It is required to machine a section 250 × 250 square for a depth of 8 mm.

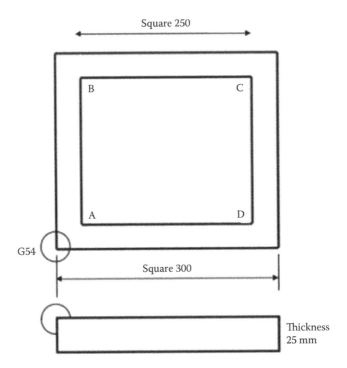

Square 250

Square 300

Thickness
25 mm

G54

B   C

A   D

FIGURE E17.12
The section to be machined from the billet.

(a) Assume work offset and the tool length compensation, (b) determine the coordinates of the corner points: A, B, C, and D.

Solution

In Figure E17.12, G54 shows the tool's original position with the coordinate point (x, y): $(0,0)$.

(a) The assumptions are as follows: work offset: G54; tool length compensation: H01.
(b) The x-coordinate of $A = (300 - 250)/2 = 25$; the y-coordinate of $A = (300 - 250)/2 = 25$.

Point A: $X = 25$ and $Y = 25$;   Point B: $X = 25$ and $Y = 275$
Point C: $X = 275$ and $Y = 275$;   Point D: $X = 275$ and $Y = 25$

EXAMPLE 17.13: PART PROGRAMMING FOR CAM/CNC MACHINING

By using the data and the drawing in Example 17.12, write a part program for CNC machining of the section from the billet.

Solution

The part program for CNC machining is presented as follows:

P101	Program Number
N10 G21 G94	Metric units input, feed rate in mm/min.
N20 G91 G28 X0 Y0 Z0	The tool is at original position.
N30 T01 M06	The tool is 17 mm diameter; end drill.
N40 G90 G54 G00 X25.0 Y25.0	Work offset call.
N50 G43 H01	Tool length compensation call.
N60 M03 S800 Z25.0	Position the tool above the starting point: *A*. The spindle rotates at 800 rpm.
N70 G01 Z–8.0 F350	The tool is linearly moved inside the work to a depth of 8 mm.
N80 G01 X25.0 Y275.0	The tool is moved to the corner point *B*.
N90 G01 X275.0	The tool is moved to the corner point *C*.
N100 G01 Y25.0	The tool is moved to the corner point *D*.
N110 G01 X25.0	The tool is moved to the corner point *A*.
N120 G00 Z100.0 M05	The tool is taken out of work; spindle stops.
N130 G91 G28 X0 Y0 Z0	Tool is moved back to the original position.
N140 M30	End of program.

Questions and Problems

17.1 Define *CAM*. List the main activities in CAM.

17.2 Explain the role of CNC machines in CAM.

17.3 Define the term CAD. List the functions of CAD.

17.4 Explain the *graphic shape generation* function of CAD with the aid of a diagram.

17.5 Draw a diagram showing the working principle of a CNC system.

17.6 Define the following terms used in CAM: (a) part program, (b) machine control unit, and (c) processing equipment of a CNC machine.

17.7 Differentiate between open-loop and closed-loop CNC systems.

17.8 Name some software packages for CAM.

P17.9 A line, defined by coordinates of end points (1, 4) and (2, 5) and matrix *X*, is required to be translated by four units in *x*- and two units in *y*-directions. Perform the indicated operation.

P17.10 The line specified in P17.9 is required to be rotated about the origin by an angle of 35°. Perform the indicated rotational operation for CAD function.

P17.11 The worktable of an open-loop CNC system is driven by a ball screw that has 4.8 mm pitch. The table is connected to the output shaft of a stepper motor (having 40 step angles) through a gearbox whose gear ratio is 4:1. It is required to move the worktable a distance of 140 mm from its present position at table travel speed of 190 mm/min. Calculate the number of pulses required.

P17.12 By using the data in P17.11, calculate the rotational speed of the stepping motor.

P17.13 By using the data in P17.11, calculate the pulse train frequency for the CNC system.

P17.14 The ball screw of a closed-loop CNC system has a pitch of 5.5 mm and is coupled to a servomotor with a gear ratio of 4:1. The MCU of the CNC machine issues 130 pulses per revolution of the ball screw. The worktable is programmed to move a distance of 110 mm at a feed rate of 435 mm/min. Calculate the number of pulses sensed by the count comparator to verify the exact 110 mm movement of the worktable.

P17.15 The cutter head of a CNC processing equipment is located at $X = 4$ and $Y = 1$. It is desired to move the cutting tool in a clockwise circular direction defined by $X = 5.8$, $Y = 7.3$ with radius $= 12$ units at a feed rate of 130 mm/min. Write the syntax for the part program.

P17.16 Repeat Examples 17.12 and 17.13 for the billet dimensions ($200 \times 200 \times 20$) and the section dimensions (150×150) for a depth of 5 mm.

P17.17 The cutter head (Tool 5) of a CNC machine is located at $X = 8$, $Y = 2$, $Z = 1$. It is required to move the tool to $X = 17$, $Y = 8.2$, $Z = 4$ at a feed rate of 130 mm/min while coolant is OFF. The spindle is at 700 rpm. Write the *block* (serial 4) for the part program.

References

Fahad, M. Huda, Z. 2017. Computer Integrated Manufacturing. In: Z. Huda (Ed.); *Materials Processing for Engineering Manufacture*. 325–351, Zurich, Switzerland: Trans Tech.

Manton, M. Weidinger, D. 2010. *Computer Numerical Control Workbook—Generic Lathe*. Kitchener, Ontario, Canada: CamInstructor.

Overby, A. 2010. *CNC Machining Handbook: Building, Programming, and Implementation*. New York, NY: McGraw-Hill.

Qureshi, M.N. 2017. Personal Communication with Head of Department, Mechanical Engineering, DHA Suffa University, Pakistan.

Smid, P. 2007. *CNC Programming Handbook*, 3rd Edition. South Norwalk, CT: Industrial Press.

Section III

Quality and Economics of Manufacturing

18

Quality Control in Manufacturing

18.1 Quality, Quality Control, Quality Assurance, Quality Management, Quality Management System, and Cost of Quality

Quality: Quality of a product refers to conformance to requirements for customer satisfaction.

Quality Control (QC): QC is a set of processes that ensures customers receive products free from defects and meet their requirements. QC involves thorough examination and testing of the quality of products to find defects by using various inspection and statistical tools.

Quality Assurance (QA): QA involves the application of various techniques and tools to assure that the manufacturing process will produce high-quality products as per customers' satisfaction.

Quality Management (QM): QM involves the management of activities that are required to maintain a desired level of quality in an organization. These activities include setting a *quality policy*, implementing QC and QA procedures, and quality improvement (Pyzdek and Keller, 2013).

Quality Management System (QMS): QMS is a formalized system that documents procedures, processes, and responsibilities for achieving quality objectives and policies. International standards (e.g., ISO 9001:2015) define the conditions for QMS (Qureshi and Huda, 2017).

Cost of Quality (CoQ): CoQ refers to the costs of nonconformance. It is a methodology to determine the extent to which an organization's resources are used for activities that prevent poor quality, that appraise the quality of the products, and that result from failures.

18.2 The Eight Dimensions of Quality for Manufacturing

The quality of a product is said to be satisfactory if it meets the expectations of the customer. In order to achieve the objective of satisfactory quality product, the following eight dimensions of quality for manufacturing must be considered: (1) performance, (2) features, (3) reliability, (4) conformance, (5) durability, (6) serviceability, (7) aesthetics, and (8) perception.

Performance: The performance of a product is the primary consideration in manufacturing. In order to verify the performance of a product, the manufacturer must ensure that the product does what it is supposed to do, within its defined tolerances.

Features: Features of a product may be referred to as the characteristics that complement the primary operating functions. In order to verify the features, the manufacturer should ensure that the product does possess all of the features specified or required for its intended purpose.

Reliability: This dimension refers to the probability of a product not failing while performing its function. *Reliability* may be verified by ensuring that the product will consistently perform within specifications. For example, surface hardness should be consistent throughout the surface.

Conformance: A product manufactured with overlooking the dimension of conformance is ruled to be defective. In order to verify *conformance*, the manufacturer must ensure that the product does conform to the specification, which may include size, shape, properties, and so on.

Durability: Durability is a measure of a product's life. The manufacturer must ensure that the product performs well under the specified operating conditions during its service life. For example, the drive shaft of an automobile must be fatigue tested for ensuring its *fatigue life*.

Serviceability: This dimension refers to the adjustments or repairs that a product may demand when put to use. Sometimes poor serviceability may force the customer to replace the product. Hence, the manufacturer must ensure that the product is relatively easy to maintain and repair.

Aesthetics: It is a qualitative aspect of a product that reflects an individual's preference. It refers to how the product looks and feels to the customer. For example, surface brightness and attractive color generally contribute significantly toward good quality of a product.

Perception: Perception refers to the product's image in public resulting from advertisements, brand name, and peer approval. Sometimes, a high-quality product may get a bad reputation based on its poor installation. Hence, it is important to ensure that the product is satisfactorily installed.

18.3 Statistical Quality Control in Manufacturing

In manufacturing, statistical quality control (SQC) refers to the use of statistical methods in the monitoring and maintaining of the quality of products. SQC can be classified into three broad categories: (1) descriptive statistics, (2) statistical process control, and (3) acceptance sampling.

Descriptive statistics: These are the set of techniques that are used to describe quality characteristics and their relationships, which includes statistical measures such as the mean, the standard deviation, the range, the distribution of data, and the like.

Statistical process control: SPC is an industry-standard methodology for measuring and controlling quality during the manufacturing process. It involves inspecting a random sample of the output from a process in real time during manufacturing (Wheeler, 2010). The data so obtained are then plotted on a graph with predetermined *control limits*, which are determined by the capability of the process. The seven basic SPC tools are discussed in Section 18.5.

Acceptance sampling: This is the process of randomly inspecting a sample of the outputs and deciding whether to accept the entire lot or reject it (see Section 18.4).

18.4 Acceptance Sampling and Probability Distributions in SQC

In acceptance sampling, SQC data are analyzed and conclusions are drawn while considering all possible variations. The SQC technique uses various tools to observe, detect, and handle possible variations and causes of the variations in products; one of the tools is *probability distribution*.

A *probability distribution* is a mathematical model that describes the value of a *random variable* with the chances of occurrence of that value in the population. A *random variable* is the variable that is subject to variation due to random causes. A probability distribution may be either a *discrete probability distribution* or a *continuous probability distribution*. In discrete distribution, random variables being measured are assigned any value from discrete value as integers. In general, discrete probability distributions are expressed by using the following mathematical models: the *binomial distribution*, the *Poisson distribution*, and the *hypergeometric distribution*. The continuous probability distributions include *normal distribution*, *exponential distribution*, and *gamma distribution*.

Binomial Distribution: Consider a process that involves a sequence of n independent trials. The outcome of each trial is independent of the outcome of any other trial; the outcome of each trial is either a "success" or a "failure." The probability that a random variable X with binomial distribution $B(n, p)$ is equal to the value k, is given by

$$P(X) = \binom{n}{k} p^k (1-p)^{n-k} = {}_nC_k p^k (1-p)^{n-k}; k = 0, 1, 2, \ldots, n \qquad (18.1)$$

where n is the number of trials, x is the number of successful trials, and p is the probability of success.

The mean of binomial distribution, μ, is given by

$$\mu_{bd} = n \cdot p \qquad (18.2)$$

The variance of binomial distribution, $(\sigma_{bd})^2$, is given by

$$(\sigma_{bd})^2 = n \cdot p \,(1 - p) \tag{18.3}$$

Poisson Distribution: The *Poisson distribution* is used to calculate the probability of an event occurring within a given time interval. The Poisson distribution with parameter $\lambda > 0$ is given by

$$P(X) = (e^{-\lambda} \lambda^k)/k! \quad k = 0, 1, 2, 3 \tag{18.4}$$

Hypergeometric Distribution: This is the probability distribution of a hypergeometric random variable, k. Let the sample population be N. The hypergeometric distribution is used to model the probability of getting k successes in n draws without replacement. It is given by

$$P(X) = \left[\binom{K}{k} \binom{N-K}{n-k} \right] \Big/ \binom{N}{n}; \quad k = 0, 1, 2, \ldots \tag{18.5}$$

The mean in hypergeometric distribution, μ_{hgd}, is given by

$$\mu_{hgd} = (n\,k)/N \tag{18.6}$$

The variance in hypergeometric distribution, $(\sigma_{hgd})^2$, is given by

$$(\sigma_{hgd})^2 = [(n\,k)\,(N-k)\,(N-n)]/[N^2\,(N-1)] \tag{18.7}$$

The significances of Equations 18.5 through 18.7 are illustrated in Example 18.3 through Example 18.5.

Exponential Distribution: Exponential distribution finds wide applications in reliability engineering since it involves a constant failure rate. This distribution has been used to mathematically model the lifetime of electronic and electrical components. If the time to failure follows an exponential behavior, the reliability function is given by

$$R(t) = e^{-\lambda t} \tag{18.8}$$

where λ is the constant failure rate, and t is the time to failure (see Example 18.6).

Gamma Distribution: The *gamma distribution* can be thought of as a waiting time between *Poisson distributed* events. It is a distribution that arises naturally in processes for which the waiting times between events are relevant. The *gamma distribution* can assume many different shapes depending on the values of the shape parameter (α) and the scale parameter (β). The mathematical expression for the failure density function for a gamma distribution is given elsewhere (Melchers, 1999). The mean of the gamma distribution ($\mu_{G.d}$) is given by

$$\mu_{G.d} = \alpha/\beta \tag{18.9}$$

The variance of the gamma distribution, $(\sigma_{G.d})^2$, is calculated by

$$(\sigma_{G.d})^2 = \alpha/\beta^2 \tag{18.10}$$

The significance of Equation 18.9 is illustrated in Example 18.7.

18.5 Statistical Process Control

18.5.1 The Seven Basic SPC Tools

An introduction to SPC is given in the preceding section. SPC involves the application of powerful problem-solving tools that are useful in achieving process stability and improving capability through the reduction of variability. There are seven basic SPC tools: (1) histogram, (2) check sheet, (3) Pareto chart, (4) flowchart, (5) cause-and-effect diagram, (6) scatter diagram, and (7) control chart. These tools are discussed in the following paragraphs.

Histogram: A histogram is a graphical representation of the distribution of numerical data using bars of different heights (Sullivan, 2010). Figure 18.1 illustrates the histogram for the tensile strength data for a polymeric material.

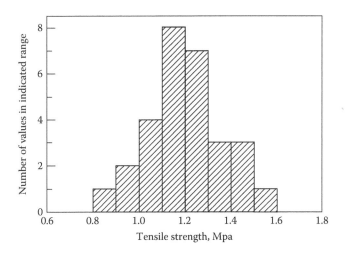

FIGURE 18.1
Histogram for tensile strength data for a polymer.

Check Sheet: A *check sheet* is a structured and prepared document used for collecting and analyzing data. The check sheet is typically a blank form that is designed for the quick, easy, and efficient recording of the desired information, which may be either quantitative or qualitative.

Pareto Chart: A *Pareto chart* analyzes the frequency of problems or causes in a process. It contains both bars and a line graph, where individual values are represented in descending order by bars, and the cumulative total is represented by the line.

Cause-and-Effect Diagram: A *cause-and-effect diagram* is a quality tool that is used to identify root causes of a problem or a bottleneck in the process and identify the reason of process malfunctioning. A list of possible solutions is gathered to draw logical conclusions (Huda, 2005).

Scatter Diagram: A *scatter diagram* is a graphical plot using Cartesian coordinates. The pattern of the resulting points (scatter) reveals any correlation present.

Control Chart: A *control chart* is a graphical quality tool that is used to analyze process variations over time. It is discussed in detail in Section 18.5.2.

18.5.2 Control Charts in SPC

A control chart comprises three horizontal lines: (1) center line (CL) that indicates the process average, (2) upper control limit (UCL), and (3) lower control limit (LCL) (see Figure 18.2). The three lines are determined from historical process data. Once data are obtained for a current process, conclusions can be drawn by comparing current data to the three lines (Figure 18.2) about the process variation—that is, whether the variation is consistent (in control) or is unpredictable (out of control, affected by special causes of variation).

There are two main types of control charts: (1) control charts for variables ($\bar{X}$ and R charts) and (2) control charts for attributes; the $\bar{X}$ and R charts are discussed in the following paragraphs.

$\bar{X}$ *chart and R chart* are the graphical representations of variations of a process over a period of time. The $\bar{X}$ *chart* (or *top chart*) monitors the average, or the central tendency of the distribution of data from the process. The

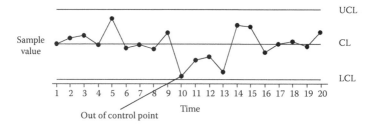

FIGURE 18.2
A control chart for variables.

quantity $\bar{x}$ is the average for each sample size inspected. The CL in Figure 18.2 is the average of averages ($\bar{\bar{x}}$) and represents the population mean.

The *R chart* (or *bottom* chart) monitors the range (R), or the width of the data distribution. Here, the CL in Figure 18.1 would represent $\bar{R}$—the average of R values.

For $\bar{X}$ *chart*, the UCL and the LCL are given by

$$\mathrm{UCL}_{\bar{x}} = \bar{\bar{x}} + A_2\bar{R} \tag{18.11}$$

$$\mathrm{LCL}_{\bar{x}} = \bar{\bar{x}} - A_2\bar{R} \tag{18.12}$$

where A_2 is the constant in the Constants for Control-Charts Table (Table 18.1).

For R charts, the UCL and LCL are given by (Kalpakjian and Schmid, 2006)

$$\mathrm{UCL}_R = D_4\bar{R} \tag{18.13}$$

$$\mathrm{LCL}_R = D_3\bar{R} \tag{18.14}$$

where the constants D_4 and D_3 are taken from Table 18.1, which also shows the values of d_2.

For acceptable quality, all parts must be within three standard deviations of the mean ($\pm 3\sigma$). The standard deviation (σ) can be calculated by (Patnaik, 1948)

$$\sigma = \bar{R}/d_2 \tag{18.15}$$

The significance of Equations 18.11 through 18.15 is illustrated in Examples 18.8 through 18.12.

TABLE 18.1

Control-Charts Constants Table

Sample Size, n	A_2 (X Chart Constant)	d_2 (for Sigma Estimation)	D_3 (R Chart Constant)	D_4 (R Chart Constant)
2	1.880	1.128	0	3.267
3	1.023	1.693	0	2.575
4	0.729	2.059	0	2.282
5	0.577	2.326	0	2.115
6	0.483	2.534	0	2.004
7	0.419	2.704	0.076	1.924
8	0.373	2.847	0.136	1.864
9	0.337	2.970	0.184	1.816
10	0.308	3.078	0.223	1.777

18.6 Six Sigma

The concept of the *Six Sigma* quality standard was first introduced by Motorola in the 1980s. Six Sigma (6σ) refers to the range of the distribution from the manufacturing process. In order to achieve *Six Sigma*, a process must not produce more than 3.4 defects per million opportunities. Hence, *Six Sigma* is a measure of quality that strives for near perfection.

18.7 Statistical Tolerancing

Statistical tolerancing (ST) is a QA technique applied to mass production. It involves taking the variation of a set of inputs to calculate the expected variation of an output of interest. ST is used to manage variation in mechanical assemblies or systems. The *ST analysis* uses a root sum square approach to develop assembly tolerances based on component tolerances; the assembly tolerance (T_{assy}) is calculated by using the following formula (Groover, 2015):

$$T_{assy} = \sqrt{\Sigma T_i^2} \tag{18.16}$$

where T_i is the tolerance of a component i in the assembly (see Example 18.13).

18.8 Problems and Solutions—Examples in QC in Manufacturing

EXAMPLE 18.1: CALCULATING THE PROBABILITY OF PRODUCTION OF DEFECTIVE UNITS IN A LOT

A company's production department determined that 25% of the units produced daily are defective. The department plans to produce 10 units during 1 hour. Calculate the probability that (a) three units would be defective and (b) at least one unit would be defective.

Solution

(a) $n = 10$, $k = 3$, $p = 25\% = 0.25$.

By using Equation 18.1,

$$P(3) = {}_nC_k p^k (1 - p)^{n-k} = {}_{10}C_3 (0.25)^3 (1 - 0.25)^{10-3}$$
$$= \left(\frac{10!}{3!(10 - 3)!} \right) (0.25)^3 (0.75)^7$$

$$P(3) = \left(\frac{10 \times 9 \times 8 \times 7 \times 6 \times 5 \times 4 \times 3 \times 2 \times 1}{(3 \times 2 \times 1)(7 \times 6 \times 5 \times 4 \times 3 \times 2 \times 1)}\right)(0.25)^3(0.75)^7$$
$$= 120 \times 0.0156 \times 0.1333 = 0.25$$

There is a 25% probability that three units would be defective out of 10 units produced in 1 hour.

(b) First we find the probability that no unit would be defective. $n = 10$, $p = 0.25$ and $k = 0$.

By using Equation 18.1,

$$P(0) = {}_{10}C_0 \,(0.25)^0 \,(1 - 0.25)^{10-0} = (1)\,(1)\,(0.056) = 0.056$$

The probability that at least one unit would be defective $= P(1 \leq x \leq 10) = 1 - P(0) = 1 - 0.056 = 0.944$

There is 94.4% probability that one unit out of 10 units produced in 1 hour would be defective.

EXAMPLE 18.2: CALCULATING THE MEAN AND VARIANCE OF BINOMIAL DISTRIBUTION FOR SQC

A company's production department determined that 32% of the units produced daily are defective. The department plans to produce 10 units during 1 hour. Calculate the following: (a) the mean of binomial distribution and (b) the variance of binomial distribution.

Solution

(a) $n = 10$, $p = 32\% = 0.32$, $\mu = ?$

By using Equation 18.2,

$$\mu = n \cdot p = 10 \times 0.32 = 3.2$$

The mean of binomial distribution $= 3.2$.

(b) By using Equation 18.3,

$$\sigma^2 = n \cdot p\,(1 - p) = (10 \times 0.32)\,(1 - 0.32) = 3.2 \times 0.68 = 2.176$$

The variance of binomial distribution $= 2.176$.

EXAMPLE 18.3: COMPUTING PROBABILITY OF PRODUCING DEFECTIVE UNITS—HYPERGEOMETRIC DISTRIBUTION

In a production department, a batch of 15 cups is found to contain 6 defective cups. The samples of size 4 are drawn without replacement, from the batch of 15. Compute the probability that a sample contains three defective cups.

Solution

Sample population $= N = 15$, sample size $= n = 4$, number of defective cups: $K = 6$, $k = 3$.

This problem involves the hypergeometric distribution, to which Equation 18.5 is applicable.

$$\binom{K}{k} = {}_{K}C_{k} = {}_{6}C_{3} = 20$$

$$\binom{N-K}{n-k} = {}_{(N-K)}C_{(n-k)} = {}_{(15-6)}C_{(4-3)} = {}_{9}C_{1} = 9$$

$$\binom{N}{n} = {}_{N}C_{n} = {}_{15}C_{4} = 910$$

By using Equation 18.5,

$$P(X) = \left[\binom{K}{k}\binom{N-K}{n-k}\right] \Big/ \binom{N}{n} = (20 \times 9)/910 == 0.197 \cong 0.2 = 20\%$$

Hence, 20% is the probability that in the sample size of four cups, three cups would be defective.

EXAMPLE 18.4: CALCULATING THE MEAN IN THE HYPERGEOMETRIC DISTRIBUTION

By using the data in Example 18.3, calculate the mean in the hypergeometric distribution.

Solution

By using Equation 18.6,

$$\mu_{hgd} = (n\,k)/N = (4 \times 3)/15 = 0.8$$

The mean in the hypergeometric distribution $= 0.8$.

EXAMPLE 18.5: CALCULATING THE VARIANCE IN THE HYPERGEOMETRIC DISTRIBUTION

By using the data in Example 18.3, calculate the variance in the hypergeometric distribution.

Solution

Sample population $= N = 15$, sample size $= n = 4$, number of defective cups: $K = 6$, $k = 3$.

By using Equation 18.7,

$$(\sigma_{hgd})^2 = [(n\,k)\,(N-k)\,(N-n)]/[N^2\,(N-1)] = [(4 \times 3)\,(15-3)$$
$$(15-4)]/[15^2\,(15-1)] = 0.5$$

The variance in the hypergeometric distribution $= 0.5$.

EXAMPLE 18.6: CALCULATING THE RELIABILITY OF A COMPONENT EXHIBITING EXPONENTIAL DISTRIBUTION

A manufacturer performs a service life test on light-bulbs and finds they exhibit constant failure rate of 0.001 failure per hour. What is the reliability of the light bulb at 15 hours of service life?

Solution

The constant failure rate $= \lambda = 0.001$ failure/h; $t = 15$ h, $R(t) = ?$

By using Equation 18.8,

$$R(t) = e^{-\lambda t} = e^{-(0.001 \times 15)} = e^{-0.015} = 0.985 = 98.5\%$$

Thus, there is 98.5% probability that the light bulb would survive (without failure) for 15 hours.

EXAMPLE 18.7: CALCULATING THE AVERAGE LIFE OF EQUIPMENT FOR RELIABILITY ANALYSIS

The service life (in hours) of equipment follows the *gamma distribution*. The shape parameter is 3.8, and scale parameter is 1/150. Calculate the average life of the equipment.

Solution

Shape parameter $= \alpha = 3.8$, scale parameter $= \beta = 1/150 = 0.0067$, average $= $ mean $= \mu = ?$

By using Equation 18.9,

$$\mu_{G.d} = \alpha/\beta = (3.8)/0.0067 = 567.1$$

The average service life of the equipment $= 567$ hours.

EXAMPLE 18.8: CALCULATING THE AVERAGE AND RANGE FOR EACH SAMPLE IN SAMPLE POPULATION IN SPC

The diameter (mm) of a machined shaft was measured for acceptance sampling; the data so obtained are presented in Table E18.8(a). The sample population is eight, and the sample size is four. Extend the Table E18.8 (a) to develop a new Table E18.8(b) by showing the following quantities

TABLE E18.8(a)

Acceptance Sampling Data for a Machined Shaft's
Diameter (mm)

Sample Number	x_1	x_2	x_3	x_4
1	6.44	6.41	6.42	6.45
2	6.42	6.43	6.38	6.44
3	6.39	6.42	6.41	6.40
4	6.41	6.42	6.43	6.44
5	6.40	6.38	6.41	6.39
6	6.45	6.44	6.46	6.47
7	6.43	6.41	6.42	6.40
8	6.38	6.40	6.39	6.41

for the sample size for each sample in the sample population: (a) average
($\bar{x}$) and (b) range (R).

Solution

For sample number 1, $\bar{x} = \dfrac{6.44 + 6.41 + 6.42 + 6.45}{4} = 6.430.$

Similarly, $\bar{x}$ for other samples can be computed, and the values are shown
in Table E18.8(b).

For sample number 1, $R = 6.45 - 6.41 = 0.04.$

The $\bar{x}$ and R data for all samples in the population are shown in
Table E18.8(b).

TABLE E18.8(b)

The $\bar{x}$ and R Data for All Samples in the Population

Sample Number	x_1	x_2	x_3	x_4	$\bar{x}$	R
1	6.44	6.41	6.42	6.45	6.430	0.04
2	6.42	6.43	6.38	6.44	6.417	0.06
3	6.39	6.42	6.44	6.40	6.412	0.05
4	6.41	6.46	6.39	6.44	6.425	0.07
5	6.37	6.38	6.43	6.39	6.392	0.06
6	6.39	6.40	6.42	6.47	6.420	0.08
7	6.43	6.41	6.42	6.40	6.415	0.03
8	6.38	6.40	6.39	6.41	6.395	0.03

EXAMPLE 18.9: CALCULATING THE AVERAGE OF THE AVERAGES AND THE AVERAGE OF RANGE IN SPC

By using the data in Table E18.8(b), calculate the (a) average of the averages ($\bar{\bar{x}}$) and (b) $\bar{R}$.

Solution

$$\bar{\bar{x}} = \frac{6.430 + 6.417 + 6.412 + 6.425 + 6.392 + 6.420 + 6.415 + 6.395}{8} = 6.414$$

$$\bar{R} = \frac{0.04 + 0.06 + 0.05 + 0.07 + 0.06 + 0.08 + 0.03 + 0.03}{8} = 0.052$$

EXAMPLE 18.10: CALCULATING THE UCL AND LCL FOR $\bar{X}$ CHART

By using the data in Example 18.9, calculate the UCL and LCL for $\bar{X}$ *chart*.

Solution

$\bar{\bar{x}} = 6.414$, $\bar{R} = 0.052$, $A_2 = 0.729$ for sample size, $n = 4$ (see Table 18.1).
By using Equation 18.11,

$$\text{UCL}_{\bar{x}} = \bar{\bar{x}} + A_2\bar{R} = 6.414 + (0.729 \times 0.052) = 6.414 + 0.038 = 6.452$$

By using Equation 18.12,

$$\text{LCL}_{\bar{x}} = \bar{\bar{x}} - A_2\bar{R} = 6.414 - (0.729 \times 0.052) = 6.414 - 0.038 = 6.376$$

Hence, for $\bar{X}$ *chart*, UCL = 6.452 and LCL = 6.376.

EXAMPLE 18.11: CALCULATING THE UCL AND LCL FOR R CHART

By using the data in Example 18.9, calculate the UCL and LCL for R chart.

Solution

By reference to Table 18.1, for sample size, $n = 4$, $D_3 = 0$ and $D_4 = 2.282$; $\bar{R} = 0.052$.

By using Equations 18.13 and 18.14, respectively,

$$\text{UCL}_R = D_4\bar{R} = 2.282 \times 0.052 = 0.118$$

$$\text{LCL}_R = D_3\bar{R} = 0 \times 0.052 = 0$$

Thus, for R chart UCL = 0.118 and LCL = 0.

EXAMPLE 18.12: CALCULATING THE STANDARD DEVIATION FOR THE PROCESS POPULATION FOR SPC

By using the data in Example 18.9, calculate the standard deviation for the process population.

Solution

By reference to Table 18.1, for sample size, $n = 4$, $d_2 = 2.059$; $\bar{R} = 0.052$.

By using Equation 18.15,

$$\sigma = \bar{R}/d_2 = 0.052/2.059 = 0.025$$

The standard deviation for the process population $= 0.025$.

EXAMPLE 18.13: COMPUTING A COMPONENT'S TOLERANCE IN AN ASSEMBLY—ST ANALYSIS APPROACH

A mechanical assembly is made up of three components. The overall dimension of the assembly is 120.0 ± 0.42 mm, and each component has a dimension of 40 mm. It may be assumed that all component tolerances are equal. Compute the component tolerance using ST analysis approach.

Solution

The assembly tolerance $= T_{assy} = 0.42$ mm.

The tolerance of a component $T_a = T_b = T_c = T_i$ (all three component tolerances are equal).

By using Equation 18.16,

$$T_{assy} = \sqrt{\Sigma T_i^2} = \sqrt{(T_i^2 + T_i^2 + T_i^2)} = \sqrt{3T_i^2}$$

$$0.42 = \sqrt{3T_i^2}$$

$$3T_i^2 = (0.42)^2 = 0.176$$

$$T_i = 0.242 \, \text{mm}$$

Hence, the tolerance on the individual component in the assembly $= 0.242$ mm.

Questions and Problems

18.1 Give meanings and definitions of the following acronyms: QC, QA, QM, QMS, CoQ, SQC.

18.2 List the eight dimensions of quality for manufacturing; explain each one of them.

18.3 List and define the seven basic SPC tools. Sketch $\bar{X}$ and R charts.

18.4 What is meant by Six Sigma? Which SPC technique may be called Three Sigma?

P18.5 A company's production department determined that 35% of the units produced daily are defective. The department plans to produce 15 units during 1 hour. Calculate the following: (a) the mean of binomial distribution and (b) the variance of binomial distribution.

P18.6 A company's production department determined that 30% of the units produced daily are defective. The department plans to produce 12 units during 1 hour. Calculate the probability that four units would be defective.

P18.7 A manufacturer performs a service life test on refrigerators and finds they exhibit a constant failure rate of 0.0006 failure per hour. What is the reliability of the refrigerator at 800 hours service life?

P18.8 In a production department, a batch of 20 cups is found to contain 8 defective cups. The samples of size 4 are drawn without replacement, from the batch of 20. Compute the probability that a sample contains two defective cups.

P18.9 A mechanical assembly is made up of two components. The overall dimension of the assembly is 80.0 ± 0.30 mm; and each component has a dimension of 40 mm. Assume that both component tolerances are equal. Compute the component tolerance using the ST analysis approach.

P18.10 The lengths (mm) of machined components were measured for acceptance sampling; the data so obtained are presented in Table P18.10. The sample population is 7, and the sample size is 3. Calculate the (a) UCL, (b) LCL, and (c) standard deviation.

TABLE P18.10

Acceptance Sampling Data for Lengths of a
Machined Component

Sample Number	x_1	x_2	x_3
1	9.26	9.21	9.28
2	9.22	9.20	9.19
3	9.24	9.18	9.22
4	9.29	9.25	9.23
5	9.27	9.22	9.19
6	9.25	9.20	9.26
7	9.20	9.25	9.28

References

Groover, M.P. 2015. *Fundamentals of Modern Manufacturing*, 6th Edition. New York, NY: Wiley.

Huda, Z. 2005. Reengineering of heat-treatment system for axle-shafts of racing cars. *Materials and Design*, 26(6), 561–565.

Kalpakjian, S., Schmid, S.R. 2006. *Manufacturing Engineering and Technology*. Upper Saddle River, NJ: Pearson Education.

Melchers, R.E. 1999. *Structural Reliability Analysis and Prediction*. New York, NY: Wiley-Blackwell.

Patnaik, P.B. 1948. The power function of the test for the difference between two proportions in a 2×2 table. *Biometrika*, 35, 157–173.

Pyzdek, T., Keller, P. 2013. *Handbook for Quality Management*, 2nd Edition. New York, NY: McGraw-Hill Professional.

Qureshi, S.M., Huda, Z. 2017. Quality Assurance in Manufacturing. In: *Materials Processing for Engineering Manufacture*. 407–432, Zurich, Switzerland: Trans Tech.

Sullivan, M. 2010. *Fundamentals of Statistics*, 3rd Edition. New York, NY: Pearson Education.

Wheeler, D.J. 2010. *Understanding Statistical Process Control*. Knoxville, TN: SPC Press.

19

Economics of Manufacturing

19.1 Cost Terminologies, Definitions, and Applications

19.1.1 Basic Cost Concepts

It is important for engineers to be familiar with a variety of cost concepts, since *costs* strongly influence the economic aspects of manufacturing activities and project analysis. These costs are defined in the following paragraphs.

Manufacturing Cost: This cost is directly involved in the manufacturing of a product. It is the sum of *direct materials cost, direct labor cost,* and *manufacturing overhead costs* (see Equation 19.1).

Nonmanufacturing Costs: These costs are not directly involved in the manufacturing process of a product. Examples of nonmanufacturing costs include marketing or selling costs, administrative costs, and the like.

Cost Object: It is the object about which cost data are required. Examples of *cost object* include a product, a process, an organizational subunit, and the like.

Fixed Cost (FC): It is the cost that is incurred for a particular period of time and within which a certain capacity level remains unaffected by the change in output level.

Variable Cost (VC): It is the cost that varies with the quantity of products produced. Examples of VC include raw material cost, wages of floor or line workers, and the like.

Direct Cost: It is the cost that can be easily and conveniently traced at a per unit level. For example, raw material cost is a direct cost.

Indirect Cost or Overhead (OH): It is the cost that cannot be easily and conveniently traced at a per unit level. Examples of OH include plant operating cost (e.g., heating and lighting), general repair cost, supervision cost, and the like.

Direct Materials Cost: It is the cost of direct material that can be easily identified from the product. Examples of direct materials cost include cost of wood in wooden table, cost of steel sheet in a car, and the like.

Direct Labor Cost: It is the labor cost that can be easily traced to individual units of products. Examples of direct labor costs include the labor cost of assembly line workers, machine operator's cost, and the like. The sum of all direct costs (direct material and direct labor) is the *prime cost*.

Manufacturing Overhead: It includes all costs of manufacturing, excluding direct material and direct labor costs. Examples of manufacturing overhead includes costs of indirect material, indirect labor, maintenance and repairs on production equipment, heat and light, property taxes, depreciation, and insurance on manufacturing facilities.

Indirect Materials: These are minor items such as solder and adhesives used in the manufacturing process. These are not included in direct materials costs.

Indirect Labor: It is a labor cost that cannot be traced to the creation of products, for example, the labor cost of janitors, supervisors, materials handlers, and night security guards.

Quality Costs: These are the costs that are incurred in controlling and assuring quality as well as the loss incurred when quality is not achieved (Qureshi et al., 2014). Quality costs can be classified into four groups: (1) prevention costs, (2) appraisal costs, (3) internal failure costs, and (4) external failure costs.

Incremental Cost: It is the additional cost that results from increasing the output by one or more units within the capacity.

19.1.2 Mathematical Models for Manufacturing Costs

The manufacturing cost can be calculated as follows:

$$\text{Manufacturing cost} = (\text{direct materials cost}) + (\text{direct labor cost}) + (\text{FOH}) \tag{19.1}$$

where FOH is the factory overhead.

Material costs are the cost of the raw materials *less* the value of the scrap:

$$C_m = W_o C_o - (W_o - W_f)C_s \tag{19.2}$$

where C_m is the material cost in manufacturing a unit, \$; W_o is the initial stock weight, kg; W_f is the final part weight, kg; C_o is the price/kg of the stock, \$; and C_s is the price/kg of the scrap, \$.

19.2 Cost and Revenue Estimation

The economics of manufacturing is strongly driven by decision-making analysis, which generally involves future cash inflows (e.g., revenues) and future cash outflows (e.g., costs). The important terminologies in the cost and revenue estimation include life cycle cost (LCC), break-even point, and cost estimation models. These terminologies are explained in the following sections.

19.2.1 Life Cycle Cost

LCC is the sum of all costs including direct costs, indirect costs (overheads), recurring costs, and nonrecurring costs associated with any project, product, process, equipment, or machinery during its life span. LCC analysis can be used for forecasting of cost profiles and performance trade-off against cost.

19.2.2 Break-Even Analysis

Basics: A *break-even* point is the output level at which total revenue (producer's income) received equals the total costs associated with the production (see Figure 19.1). It is the most widely applied analytical and managerial tool, which analyzes the cost-output-profits relationship at different levels of output. The break-even analysis facilitates top management in decision-making, budgeting, operational and production planning, profit forecasting, establishing and reviewing pricing policies, and controlling manufacturing costs.

Total revenues can be calculated by multiplying by the quantity of the products sold by the unit price of the product (see Equation 19.3). Break-even analysis is also helpful in calculating the *margin of safety*. During the economic downturn, the margin of safety helps managers to make the right decisions on their production and sales. The margin of safety is the total sales minus sales at the break-even point (see Equation 19.6). The mathematical model of break-even analysis is given in Section 19.2.

Mathematical Modeling: The *total revenue* is given by

$$TR = P \cdot Q \tag{19.3}$$

where TR is the total revenue, \$; P is the unit price of the good sold, \$; and Q is the quantity of goods sold.

The cost per unit of a manufactured product is given by

$$\text{Cost per unit} = (FC + VC)/Q_m \tag{19.4}$$

where FC is the fixed cost, VC is the variable cost of the output level, and Q_m is the number of units manufactured.

The break-even point is the quantity referring to no loss–no profit, and it can be expressed as

$$BE_Q = (FC)/(P-V) \tag{19.5}$$

where BE_Q is the break-even point; P is the unit selling price, \$; and V is the unit variable cost, \$. Before this point, $TR < TC$ that is a loss condition (see Figure 19.1). Every additional unit sold is used to cover the FC and is known as the contribution margin. After this output level $TR > TC$, there is a profit condition (see Figure 19.1):

$$\text{Unit contribution margin} = \text{Selling price} - \text{Variable cost} \tag{19.6}$$

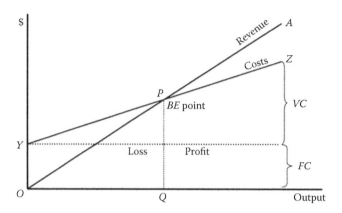

FIGURE 19.1
Break-even point (VC = variable cost; FC = fixed cost).

The margin of safety is given by

$$\text{Margin of safety} = \text{Total sales} - \text{Sales at } BE \text{ point} \qquad (19.7)$$

The significances of Equations 19.3 through 19.7 are illustrated in Examples 19.3 through 19.5.

19.2.3 Cost Estimation Models

Basics: Cost estimation models are various mathematical modeling techniques that are used to calculate future cost and revenue estimates for economic analysis. These models include cost index, unit and factor technique, power-size technique, and *learning curve* technique.

Cost Indexes: Cost indexes are used to estimate future cost (or price) of an item based on its historical data. It is a unit-less measurement and enables us to calculate the estimated price or cost of an item in the year n (see Equation 19.8).

Unit and Factor Technique: The *unit technique* is a simple model of cost estimation in which cost per unit factor, which can be projected accurately, is used to estimate the total cost. The *factor technique* involves estimation of the total cost of several quantities by adding the unit cost of every component and adding the cost of any directly estimated component (see Equation 19.9).

Power Sizing Technique: This model is based on the concept of economies of scale and enables us to calculate the cost of plant 2 when the size of the plant and the cost and size of plant 1 are given along with a power factor (cost capacity factor, x) (see Equation 19.10).

Learning Curve Technique: This technique is based on the concept that by performing a repetitive task, a worker becomes more efficient due to which input resource utilization is enhanced. This technique enables us to calculate the *learning curve exponent* (see Equation 19.11).

Mathematical Modeling: Cost index, a unit-less measurement, is used to estimate future cost (or price) of an item based on historical data. It is given by

$$C_n = C_k \, (I_n/I_k) \qquad (19.8)$$

where k is the reference year for which price or cost of an item is known, n is the year for which price or cost needs to be estimated $(n > k)$, C_n is the estimated price or cost in year n, and C_k is the price or cost of an item in reference year k. I_n and I_k are the index values in current and reference years, respectively.

The *estimated cost*, C, is given by

$$C = \sum_d C_d + \sum_m C_m Q_m \qquad (19.9)$$

where C_d is the directly estimated cost of component d, C_m is the cost per unit of component m, and Q_m is the number of units of component m.

The cost of the plant can be calculated by

$$C_2 = C_1 \left(\frac{S_2}{S_1}\right)^x \qquad (19.10)$$

where C_1 is the cost of plant 1, C_2 is the cost of plant 2, S_1 is the size of plant 1, S_2 is the size of plant 2, and x is the cost-capacity factor that reflects economies of scale $(1 \leq x \leq 1)$. It must be noted that C_1 and C_2 both are determined at a point in time for which an estimate is desired. S_1 and S_2 both must be in the same physical units. If $x < 1$, it indicates decreasing economies of scale (i.e., each additional unit of capacity costs less than the previous unit). If $x > 1$, it indicates increasing economies of scale (i.e., each additional unit of capacity costs more than the previous unit of capacity). If $x = 1$, that indicates a linear cost relationship with every unit capacity.

The learning curve exponent, n, is given by

$$n = (\log S)/(\log 2) \qquad (19.11)$$

where S is the learning rate or slope of the learning curve.

The number of input units required to produce output units U can be calculated by

$$Z_u = K U^n \qquad (19.12)$$

where Z_u is the number of input units required to produce output units U; K is the number of input units required to produce the first unit of output; and n is the learning curve exponent.

The significances of Equation 19.8 through 19.12 are illustrated in Examples 19.5 through 19.7

19.3 Capital Budgeting

Capital budgeting involves activities of generating, evaluating, selecting, and following up the company's long-term capital investment projects that are worth the funding of cash through the firm's capitalization structure. Examples of capital budgeting projects include renovation of an existing manufacturing facility, expansion of a plant, research and development (R&D) projects, and the like. The evaluation criteria of a capital budgeting project include the minimum acceptable rate of return (MARR). The scope of this book does not permit a detailed account on capital budgeting; the reader is advised to refer elsewhere (Shapiro, 2004) for an in-depth discussion on this topic.

19.4 Depreciation

Basics: *Depreciation* refers to the decrease in the value of tangible assets with time, usage, and/or obsolescence. It is important to calculate *depreciation* due to two reasons: (1) for making managerial decisions (i.e., planning and replacement decisions of equipment) and (2) for tax calculations (i.e., depreciation amount is considered as "expense" in income tax calculations) (Blank and Tarquin, 2005; Sullivan et al., 2009).

19.4.1 Some Important Depreciation Terms and Definitions

First cost/Unadjusted Basis/Basis/Cost Basis (B): Total installed cost of an asset

Depreciation amount (D_t): Decrease in value of an asset in year t

Book Value $(BV)_t$: Remaining undepreciated initial cost in year t

Recovery Period (n): Depreciation life of an asset

Market Value (MV): Amount received if asset sold in market

Salvage Value (SV): Estimated value at the end of asset's useful life

Depreciation Rate (d_t): Fraction of first cost or basis removed each year t

19.4.2 Mathematical Modeling of Depreciation

Straight Line (SL) Depreciation Method: The SL depreciation method assumes that a constant amount is depreciated each year over the depreciable (useful) life of the asset; hence, the book value decreases linearly with time.

The depreciation rate, d, can be related to recovery period, n, as follows:

$$d = (1/n) \tag{19.13}$$

The depreciation amount, D_t, can be computed as follows:

$$D_t = d\,(B - SV) \tag{19.14}$$

where B is the *first cost*, d is the depreciation rate, and SV is the salvage value. The book value can be calculated as follows:

$$(BV)_t = (BV)_{t-1} - D_t = tD_t \tag{19.15}$$

Declining Balance (DB) and Double Declining Balance (DDB) Depreciation Methods: These depreciation methods do not involve salvage value in the depreciation calculation, so that the depreciation schedule may be just equal to the salvage value or greater than or less than the salvage value.

$$D_t = dB\,(1 - d)^{t-1} \tag{19.16}$$

$$(BV)_t = B\,(1 - d)^t \tag{19.17}$$

For DB, depreciation rate, $d = 150\%$ of SL or 175% of SL. For DDB, $d = 200\%$ of SL.

Modified Accelerated Cost Recovery System (MACRS): The MACRS method of depreciation involves calculations for depreciation rates for recovery periods, which is given elsewhere (Accounting Explained, 2017). For MACRS,

$$D_t = d_t B \tag{19.18}$$

$$(BV)_t = (BV)_{t-1} - D_t \tag{19.19}$$

Example 19.8 illustrates the applications of Equations 19.13 through 19.19.

19.5 Inventory Management

19.5.1 Inventory Management Basics

The success of a manufacturing company is largely determined by its smooth and uninterrupted production and distribution operations, which are ensured by an effective *inventory management*. Any stored resource that a company maintains to satisfy their current and future needs is called *inventory*. The main functions of inventory management and control include (a) decoupling the production process, (b) stocking resources, (c) providing a quantity discount advantage, and (d) hedging against inflation (Heizer et al., 2004).

Inventory Models: These are mathematical models that assist in answering two questions: (1) how much should be ordered? and (2) when should it be ordered? There are two types of inventory models: (1) *deterministic inventory models*—with known and constant demand, and (2) *stochastic inventory models*—with unknown and probabilistic demand.

There are some features of inventory management that involve uncertainty, but *deterministic models* have the advantage of providing a base on which to incorporate assumptions concerning uncertainty.

Inventory Costs: In order to make a decision that affects inventory size, the following four inventory costs must be considered: holding cost, setup cost, ordering cost, and shortage cost. The *holding cost* refers to the cost of holding the inventory; it includes the costs of storage facilities, handling, insurance, breakage, obsolescence, depreciation, taxes, and the opportunity cost of capital. In case holding costs are high, it is important to keep low inventory levels and ensure frequent replacement. *Setup cost* refers to the cost incurred in setting up facilities for manufacturing; it involves procurement of necessary materials, equipment setups, documentation, charging time, and moving out the previous stock of material (if the production of different products is involved). *Ordering cost* involves clerical costs to prepare the purchase or production order; it includes all the details, such as counting items and calculating order quantities. *Shortage cost* is to the cost of stock-out (i.e., the situation when a customer seeks a product but it is not available or is out of stock). The mathematical modeling of inventory costs is given in Section 19.5.2.

19.5.2 Mathematical Modeling of Inventory Management

19.5.2.1 Economic Order Quantity (EOQ) Model

In the EOQ model, the optimal order quantity, Q^*, occurs at the point where ordering cost curve and holding cost curve intersect each other (see Figure 19.2).

The mathematical relationship for the *annual setup cost* can be derived as follows (Huda, 1998):

$$\text{Annual set-up cost} = (\text{Number of orders per year}) \\ \times (\text{Setup or order cost per order}) \quad (19.20)$$

Since annual demand (D) is known and constant, the number of orders per year will be

$$\text{Number of orders per year} = (\text{Annual demand})/ \\ (\text{Number of units in each order}) = D/Q \quad (19.21)$$

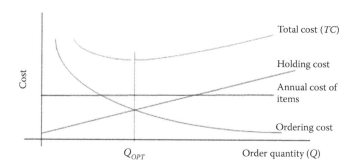

FIGURE 19.2
Total inventory cost (*TC*) as a function of order quantity (*Q*).

By combining Equations 19.19 and 19.20, we obtain

$$\text{Annual setup cost} = [D/Q] \, (S) \tag{19.22}$$

where D is annual demand, Q is the number of units in each order, and S is the setup or order cost per order.

The total annual holding cost is computed by multiplying the annual per-unit holding cost, H, times the average inventory level, $Q/2$:

$$\text{Annual holding cost} = [Q/2] \, (H) \tag{19.23}$$

We know that the optimal order quantity, Q^*, refers to the condition when the annual setup cost becomes equal to the holding cost (see Figure 19.2). Hence, the optimal order quantity can be computed by equating the right-hand sides of Equations 19.28 and 19.29, as follows:

$$[D/Q^*] \, (S) = [Q^*/2] \, (H) \tag{19.24}$$

By solving Equation 19.24 for Q^*, we obtain

$$Q^* = \sqrt{(2DS)/H} \tag{19.25}$$

The expected number of orders placed during the year (N), the expected time between the orders, and the total annual inventory cost can also be determined as follows:

$$N = D/Q^* \tag{19.26}$$

where N is the expected number of orders placed during the year, and D is the annual demand.

The expected time between orders can be computed by the following relationship:

Expected time between orders
$$= (\text{Number of working days per year})/N \qquad (19.27)$$

The total annual inventory cost, TC, is the sum of annual setup cost and annual holding cost. By using Equations 19.22 and 19.23, we can express the total annual inventory cost, TC, as follows:

$$TC = [D/Q]\,(S) + [Q/2]\,(H) \qquad (19.28)$$

The reorder point (ROP) is the inventory level at which an order must be placed. ROP depends on the *lead time*, L, which is the time of order, and assumes that lead time and demand during lead time are constant.

$$
\begin{aligned}
\text{ROP} &= (\text{Demand per day}) \times (\text{Lead time for a new order in days}) \\
&= d \times L
\end{aligned} \qquad (19.29)
$$

The demand per day, d, can be found as follows:

$$d = (D)/(\text{Number of working days in a year}) \qquad (19.30)$$

19.5.2.2 Economic Production Lot Size Model

$$\text{Number of production runs per year} = D/Q \qquad (19.31)$$

$$\text{Setup cost per year} = [D/Q]\,(S) \qquad (19.32)$$

$$
\begin{aligned}
\text{Holding cost per year} &= (\text{Average inventory level}) \\
&\quad \times (\text{Holding cost per unit per year})
\end{aligned} \qquad (19.33)
$$

$$\text{Average inventory level} = (\text{Max. inventory level})/2 \qquad (19.34)$$

$$
\begin{aligned}
\text{Max. inventory level} &= (\text{Total production during the run}) \\
&\quad - (\text{Total usage during the run})
\end{aligned} \qquad (19.35a)
$$

$$\text{Max. inventory level} = [(P)(t)] - [(d)(t)] \qquad (19.35b)$$

$$\text{Total production during production run} = (P)\,(t) = Q \qquad (19.36a)$$

$$t = Q/P \qquad (19.36b)$$

By substituting the value of t in Equation 19.35b, we get

$$
\begin{aligned}
\text{Max. inventory level} &= P\,(Q/P) - d(Q/P) = Q - d(Q/P) \\
&= Q\,[1 - (d/P)]
\end{aligned} \qquad (19.37)
$$

$$
\begin{aligned}
\text{Holding cost per year} &= [(\text{Max. inventory level})/2]\,(H) \\
&= (HQ/2)\,[1 - (d/P)]
\end{aligned} \qquad (19.38)
$$

The optimal production lot size can be determined by setting ordering (setup) cost equal to holding cost (i.e., Holding cost per year = Setup cost per year).

By reference to Equations 19.31 and 19.37, we get

$$(HQ'/2)\,[1 - (d/P)] = [D/Q']\,(S) \tag{19.39}$$

By solving Equation 19.38 for Q', we obtain

$$Q' = (\sqrt{(2DS)/H})(\sqrt{P/(P - D)}) \tag{19.40}$$

where Q' is the optimal production lot size.

The total annual cost, TC, can be computed as follows:

$$TC = \frac{D}{Q}S + \frac{Q}{2}H\left(1 - \frac{d}{p}\right) \tag{19.41}$$

Examples 19.9 through 19.10 illustrate the significance of Equations 19.20 through 19.41.

19.6 Economics of Special Tooling and Its Mathematical Modeling

For a special tooling (e.g., a jig, fixture, etc.) to be economically feasible,

$$\text{Tooling cost per piece} < (a - b) \tag{19.42}$$

where a is the total cost per piece without tooling, and b is the total cost per piece using the tooling.

In practical situations, Equation 19.42 can be expressed as

$$\{C_t + [(C_t/2)\,(n)\,(i)]\}/N < (L_t + M_t) - (L_t t_t + M t_t) \tag{19.43}$$

where
C_t = cost of special tooling;
n = number of years tooling will be used;
i = interest or profit rate invested capital is worth;
N = number of pieces that are to be manufactured using the tooling;
L = labor wage rate per hour without use of the tooling;
L_t = labor wage rate per hour using the tooling;
t = time in hours taken to manufacture a piece without tooling;
t_t = time in hours taken to manufacture a piece using the tooling; and
M = machine cost per hour, including all overhead.

The significances of Equations 19.42 through 19.43 are illustrated in Example 19.11.

19.7 Problems and Solutions—Examples on Manufacturing Economics

EXAMPLE 19.1: COMPUTING ANNUAL MANUFACTURING COST

ABC manufacturing company spent $9000 on direct materials during the year ending December 31, 2011. The direct labor cost and the factory overhead for the year were estimated to be $60,000 and $100,000, respectively. The cost of opening work in progress was $3000, and the cost of closing work in progress was $2000. Calculate the manufacturing cost for the year.

Solution

By using Equation 19.1,

$$\text{Manufacturing cost} = (\text{direct materials cost}) + (\text{direct labor cost}) + (FOH)$$
$$= (9000) + (60,000) + (100,000) = \$169,000$$

$$\text{Net cost on work in progress} = 3000 - 2000 = \$1000$$

$$\text{Actual manufacturing cost} = 169,000 + 1000 = 170,000$$

$$\text{Actual manufacturing cost during the year} = \$170,000$$

EXAMPLE 19.2: COMPUTING MATERIAL COST FOR A PRODUCT

A metal process industry purchased aluminum at $13/kg and used 2 kg of aluminum in manufacturing a cooking utensil. The final part weight was measured to be 1.7 kg. Compute the direct material cost in manufacturing a unit of the utensil, if the price of scrap is $3/kg.

Solution

Initial stock weight = W_o = 2 kg, final part weight = W_f = 1.7 kg, price of stock = C_o = $13/kg, price of scrap = C_s = $3/kg.

By using Equation 19.2

$$C_m = W_o C_o - (W_o - W_f)C_s = (2 \times 13) - [(2.0 - 1.7)3]$$
$$= (26) - (0.9) = 25.1$$

The direct material cost in manufacturing a (unit) utensil = $25.10.

EXAMPLE 19.3: COMPUTING THE COST PER UNIT OF A PRODUCT

XYZ Company has total variable costs of $50,000 and total fixed costs of $30,000 in April 2013, which it incurred while producing 10,000 flash-drives for computers. Calculate the cost per unit of the flash-drive.

Solution

$FC = \$30,000$; $VC = \$50,000$, number of units manufactured $= 10,000$.

By using Equation 19.4,

$$\text{Cost per unit} = (\$30,000 + 50,000)/10,000 \text{ units} = \$8$$

EXAMPLE 19.4: COMPUTATION OF BREAK-EVEN POINT AND MARGIN OF SAFETY

For a manufacturing plant, the fixed cost per month is \$95,000. It can produce 10,000 units per month. Total variable cost per unit (including direct and indirect costs) is \$6, and unit sales price is \$12. The estimated sales of the plant in 1 month are \$480,000. Calculate (a) the per month break-even output level of the firm and (b) the margin of safety.

Solution

(a) By using Equation 19.5,

$$BE(Q) = (FC)/(P - V) = (95,000)/(12 - 6) = (95,000)/6 = 15,834 \text{ units}$$

Hence, the per month break-even output level is 15,834 units.

(b) By using Equations 19.3 and 19.7,

$$\begin{aligned}
\text{Margin of safety} &= \text{Total sales} - \text{Sales at break-even point} \\
&= (480,000) - (15,834 \times 12) = (480,000) - (190,000) \\
&= 290,000
\end{aligned}$$

The margin of safety is \$290,000.

EXAMPLE 19.5: COMPUTING FACILITY COST

An automated molding machine delivered to a facility costs \$18 million. Installation and delivery charges are estimated to be \$0.25 M and \$0.45 M, respectively. The construction cost of the facility is \$1000/m². What would be the total development cost of a 10,000 m² facility?

Solution

$$\text{Total cost of facility} = \text{equipment price} + \text{installation cost} + \text{delivery cost}$$

$$\Sigma\, C_d = \$18\,M + \$0.25\,M + \$0.45\,M = \$18.7\,M$$

$$\Sigma\, C_m\, Q_m = (\$1000)\,(10,000) = \$10\,M$$

By using Equation 19.9,

$$\text{Total cost of facility} = C = \$18.7\,M + \$10.0\,M = \$28.7 \text{ million.}$$

EXAMPLE 19.6: COMPUTING COST OF PLANT AT PRESENT

Five years ago the cost of a 100-HP air compressor was $4700, and the cost index was 150. What is the cost of the 400-HP compressor today at 250 cost index? The cost-capacity factor of 400-HP compressor is 0.7.

Solution

The given data are as follows:

Cost of plant $1 = C_1 = \$4700$, size of plant $1 = S_1 = 100$-HP, size of plant $2 = S_2 = 400$-HP.

As C_1 and C_2 must be at a same time point, we use the cost index (given in the example) to convert C_1 at a time point of C_2.

By reference to Equations 19.8 and 19.10,

$$C_2 = 4700 \ [(400/100)^{0.7}] \ (250/150) = 4700 \times 2.639 \times 1.67 = \$20{,}672$$

Hence, the cost of a 400-HP compressor at present $= \$20{,}672$.

EXAMPLE 19.7: COMPUTING THE TIME TO MANUFACTURE A GIVEN UNIT OF A PRODUCT

An assembly-line worker required 1.8 hour to assemble the first unit of a product. If the learning rate is 85%, how many hours are required to assemble the ninth unit of the product?

Solution

Learning rate $= S = 85\% = 0.85$, $K = 1.8$ h, $U = 9$.

By using Equation 19.11,

$$n = (\log S)/(\log 2) = (\log 0.85)/(\log 2) = (-0.0706)/(0.301) = -0.2344$$

$n = -0.3189$

By using Equation 19.12,

$$Z_u = KU^n = (1.8) \ (9)^{-0.2344} = (1.8)/(9)^{0.2344} = (1.8)/(1.673) = 1.076 \text{ hour}$$

Hence, the time required to assemble the ninth unit of the product is 1.07 hour.

EXAMPLE 19.8: COMPUTING DEPRECIATION AMOUNT AND BOOK VALUE

A petroleum exploration company recently purchased a drilling machine at the cost of $75,000 with estimated market value of $14,000 at the end of 15 years of depreciable life. Calculate the fifth and the seventh year depreciation amounts and book values using (a) SL and (b) DDB depreciation methods.

Solution

$B = \$75{,}000$, $(SV) = \$14{,}000$ $n = 15$

(a) SL depreciation:

By using Equation 19.13,

$$d = (1/n) = (1/15) = 0.067$$

By using Equation 19.14,

$$D_t = d \ (B - SV) = (0.067) \ (75{,}000 - 14{,}000) = \$4087$$

By using Equation 19.15,

$$(BV)_5 = t \ D_t = 5 \times 4087 = \$20{,}435$$

$$(BV)_7 = t \ D_t = 7 \times 4087 = \$28{,}609$$

(b) DDB Depreciation,

For DDB, $d = 200\%$ (d for SL) $= 2 \times (1/15) = (2/15)$

By using Equations 19.16 through 19.18,

$$D_5 = (2/15) \ (75{,}000) \ [1 - (2/15)]^4 = 0.133 \times 75{,}000 \times 0.565 = \$5636$$

$$D_7 = (2/15) \ (75{,}000) \ [1 - (2/15)]^6 = 0.133 \times 75{,}000 \times 0.424 = \$4236$$

$$(BV)_5 = (75{,}000) \ [1 - (2/15)]^5 = 75{,}000 \times 0.489 = \$36{,}741$$

$$(BV)_7 = (75{,}000) \ (0.867)^7 = \$27{,}618$$

EXAMPLE 19.9: COMPUTING OPTIMAL ORDER QUANTITY, INVENTORY COST, AND ROP

A carpet manufacturing company wants to minimize its inventory cost. The annual demand is 2200 carpets, setup or ordering cost is $21 per order with $6 per unit holding cost. The company operates 275 days per year. Calculate the (a) optimal order quantity, Q^*; (b) the number of orders per year, N; (c) the expected time between orders; (d) the total inventory cost per year, TC; and (e) the ROP if on average it takes 4 working days to deliver an order.

Solution

Annual demand $= D = 2200$ carpets, ordering cost per order $= S = \$21$, holding cost/unit $= H = \$6$.

Working days in a year $= 275$, lead time $= L = 4$ days.

(a) By using Equation 19.25,

$$Q^* = [(2DS)/H]^{1/2} = [(2 \times 2200 \times 21)/6]^{1/2} = \sqrt{(15{,}400)} = 124$$

(b) By using Equation 19.26,

$$N = D/Q^* = (2200)/(124) = 18 \text{ orders per year}$$

(c) By using Equation 19.27,

Expected time between orders = (Number of working days per year)/N
$$= (275)/(18) = 15 \text{ days}$$

(d) By using Equation 19.28,

$$TC = [D/Q^*] \, (S) + [Q^*/2] \, (H) = [2200/124](21) + [124/2](4)$$
$$= 372.58 + 248 = \$\, 620.58$$

(e) By using Equation 19.30,

$$d = (D)/(\text{Number of working days in a year}) = (2200)/(275) = 8 \text{ units}$$

By using Equation 19.29,

$$ROP = d \times L = 8 \times 4 = 32 \text{ units}$$

EXAMPLE 19.10: COMPUTING OPTIMAL PRODUCTION LOT SIZE

Assume that the carpet manufacturing company referred to in Example 19.8 has its own outlet store as their one facility. The setup cost and holding cost are $21 and $6, respectively, and the annual demand is 4800 units. The manufacturing facility operates 280 days per year; on the same days the store remains open and produces 190 units per day. Determine the following: (a) the optimal production lot size, (b) total inventory cost, and (c) the maximum inventory level.

Solution

$D = 4800$ units, $S = \$21$, $H = \$6$, Operating days per year $= 280$, $P = 190$ units per day.

(a) By using Equations 19.32 through 19.41,

$$d = (4800/280) = 17.14 \text{ units per day}$$

$$Q^* = \{[(2DS)/H] \, [(P)/(P-d)]\}^{1/2}$$

$$Q^* = \{[(2 \times 4800 \times 21)/6] \, [(190)/(190-17.14)]\}^{1/2} = 192 \text{ units}$$

Optimal production lot size = $Q^* = 192$ units

$$TC = [(D/Q^*)\,S] + \{[(Q^*/2)\,H]\,[1 - (d/P)]\}$$

$$TC = [(4800/192)\,21] + \{[(192/2)\,6]\,[1 - (17.14/190)]\} = \$1049$$
Total cost = $1049

$$\begin{aligned}
\text{Maximum inventory level} &= Q^*[1 - (d/P)] = 192[1 - (17.14/190)] \\
&= 192(1 - 0.09) = 192 \times 0.91 \\
&= 175 \text{ units (approx.)}
\end{aligned}$$

The maximum inventory level = 175 units (approx.)

EXAMPLE 19.11: DECIDING ECONOMIC FEASIBILITY OF SPECIAL TOOLING

An automotive parts industry is planning to reduce its manufacturing cost by using a jig (a special tooling). Six hundred units are to be manufactured by using the jig, which is expected to last for 3 years. The labor wage rate without the special tooling is $7/h, whereas the labor wage rate using the jig will be reduced to $4/h. The job takes 0.4 h without tooling and 0.18 h using the tooling. If the cost of the jig is $800, would it be economical to use the jig? The investment capital is worth 5% to the company, and the machine cost is $1.8/h.

Solution

Cost of special tooling = $C_t = \$800$.

Number of years tooling will be used = $n = 3$.

Interest or profit rate invested capital is worth = $i = 5\% = 0.05$.

Number of pieces that are to be manufactured using the tooling = $N = 600$.

Labor wage rate per hour without the use of tooling = $L = \$7/h$.

Labor wage rate per hour using the tooling = $L_t = \$4/h$.

Time in hours taken to manufacture a piece without tooling = $t = 0.40\,h$.

Time in hours taken to manufacture a piece using the tooling = $t_t = 0.18\,h$.

Machine cost per hour, including all overheads = $M = \$1.8/h$.

By using the left-hand side of Equation 19.43,

$$\{C_t + [(C_t/2)(n)(i)]\}/N = \{800 + [(400)(3)(0.05)]\}/600 = 860/600 = \$1.43$$

By using the right-hand side of Equation 19.43,

$$(L_t + M_t) - (L_t t_t + Mt_t) = [(7 \times 0.4) + (1.8 \times 0.4)]$$
$$- [(4 \times 0.18) + (1.8 \times 0.18)] = \$2.48$$

Since $\{C_t + [(C_t/2)(n)(i)]\}/N < (L_t + M_t) - (L_t t_t + Mt_t)$, the use of the jig is economically feasible.

Questions and Problems

19.1 Which of the following statements are TRUE or FLASE?

(a) The sum of direct material cost and direct labor cost is the *prime cost*.

(b) The wage of a production supervisor is an example of direct labor cost.

(c) The cost of oxygen gas used in oxyfuel welding is a direct material cost.

(d) All costs of manufacturing, excluding direct material and direct labor costs, are called manufacturing overheads.

(e) The optimal order quantity occurs at the point where ordering cost and total cost curves intersect.

(f) Declining balance (DB) and DDB depreciation methods involve salvage value in the calculation of depreciation.

19.2 (a) Define the terms *break-even point* and *margin of safety*. (b) How is break-even analysis useful in decision-making by top management?

19.3 (a) What are the advantages and functions of inventory management? (b) List the various inventory costs, and explain two of them.

19.4 Define the term *capital budgeting* with examples.

19.5 Differentiate between straight line and declining balance depreciation methods for calculating depreciation.

19.6 Derive an expression for the optimal order quantity in terms of annual demand, setup cost, and holding cost.

19.7 The ABC manufacturing company has a production capacity of producing 10,000 units per month. The fixed cost of the manufacturing plant per month is $100,000. Total variable cost per unit (including direct and indirect costs) is $5, and the sales price per unit is $10. The estimated revenues from sales in 1 month are $500,000. Calculate the following: (a) the break-even output level per month of the firm and (b) the margin of safety.

19.8 The PQR manufacturing company has total variable costs of $43,000 and total fixed costs of $25,000 in September 2014, which it incurred while producing 9000 gears. Calculate the cost per unit of the gear.

19.9 An automated machine delivered to a facility costs $20 million. The installation and delivery charges are estimated to be $0.3 M and $0.5 M, respectively. The construction cost of the facility is $1000/m². What would be the total development cost of a 10,000 m² facility?

19.10 The XYZ manufacturing company spent $8600 on direct materials during the year ending December 31, 2015. The direct labor costs and the factory overheads for the year were estimated to be $55,000 and $98,000, respectively. The cost of opening work in progress was $2700, and the cost of closing work in progress was $1800. Calculate the manufacturing cost for the year.

19.11 Five years ago the cost of a 100-HP motor was $5000 and the cost index was 150. What is the cost of a 500-HP motor today at a 250 cost index? The cost-capacity factor of the 500-HP motor is 0.8.

19.12 A metal forming industry purchased stainless steel at $7/kg and used 3 kg of aluminum in manufacturing a cooking utensil. The final part weight was measured to be 2.6 kg. Compute the direct material cost in manufacturing a unit of the utensil if the price of scrap is $1.5/kg.

19.13 It requires 2 hours for an assembly-line worker to assemble the first unit of a product. The learning rate is 80%. Compute the time in hours to assemble the tenth unit of the product?

19.14 An automotive parts manufacturing company purchased a heat-treatment furnace at the cost of $65,000 with estimated market value of $8000 at the end of 15 years of depreciable life. Calculate the fourth- and sixth-year deprecation amounts and book values using (a) SL and (b) DDB depreciation methods.

19.15 A steel-furniture manufacturing company wants to minimize its inventory cost by determining the optimal number of steel sheets to obtain per order. The annual demand of steel filing cabinets is 2000 units. The ordering cost is $20 per order with a $5 per unit holding cost. The company operates 255 days per year. Calculate the following: (a) optimal order quantity, Q^*; (b) number of orders per year, N; (c) expected time between orders, (d) total inventory cost per year, TC; and (e) reorder point (ROP) if on average it takes 5 working days to deliver an order.

19.16 A steel fabrication industry manufactures a range of steel furniture, including clothing lockers (C/L). The annual demand of C/L is 500 units. The company estimates that it costs $18 to set up all the equipment and paperwork for production of this product. When a production batch is running the C/L, the production rate is 25 units per day. Each C/L costs the company $80 to produce, and the holding cost is estimated to be 15% of the manufacturing cost. The factory produces 5 days a week for 50 weeks a year. Compute the (a) optimal production lot size, (b) total inventory cost, and (c) maximum inventory level.

References

Accounting Explained 2017 Internet Source: http://accountingexplained.com/financial/non-current-assets/macrs; Accessed on April 02.

Blank, L., Tarquin, A. 2005. *Engineering Economy*. New York, NY: McGraw-Hill.

Heizer, J.H., Render, B., Weiss, H.J. 2004. *Operations Management. 8*. Upper Saddle River, NJ: Pearson Education, Prentice Hall.

Huda, Z. 1998. *Production Technology and Industrial Management*. Karachi: HEM Publishers.

Shapiro, A.C. 2004. *Capital Budgeting and Investment Analysis*. Upper Saddle River, NJ: Pearson Education.

Sullivan, W.G., Wicks, E.M., Koelling, C.P. 2009. *Engineering Economy*. Upper Saddle River, NJ: Prentice Hall.

Qureshi, S.M., Majeed, S., Khalid, R. 2014. Where do we stand in cost of quality awareness? Pakistan's case. *International Journal of Quality Engineering and Technology*, 4(4), 273–289.

Answers to Selected Problems

Chapter 2

P2.5   1,954,770 J or 1.95 MJ

P2.7   3.12 min

P2.9   (a) 86 N, (b) Nil

P2.11 56 kgf

P2.13 (a) 1.39 kg, (b) 2827 rev/min, (c) 914 N

Chapter 3

P3.7   23 mm

P3.9   (a) 160 cm/s, (b) 447 cm^3/s, (c) 2.12 s

P3.11 6 cm

Chapter 4

P4.5   2.2

P4.7   29 cm/s

P4.9   8.08 cm

P4.11 8 tf (10 tf rating)

Chapter 5

P5.7   127.6 MPa

P5.9   (a) $-34.6\,^{\circ}$C, (b) 268 $^{\circ}$C

P5.11 0.44

Chapter 6

P6.9   (a) 22.8 m/min, (b) 0.42

P6.11 (a) Yes, (b) 1.98×10^6 N

P6.13 Two passes

P6.15 250.53 kN

P6.17 1.3×10^6 N

P6.19   208 kN

P6.21   (a) 5, (b) 1.6, (c) 2.52

P6.23   1882.2 MPa

P6.25   996.1 MPa

P6.27   (a) 891.2 MPa, (b) 3.38 kN

Chapter 7

P7.7   (a) 0.2 mm, (b) 139.6 mm, (c) 140 mm

P7.9   (a) 0.225 mm, (b) 10 mm, (c) 10.45 mm

P7.11   223.1 kN

P7.13   1.7 kN

Chapter 8

P8.5   150.47 MPa

P8.7   (a) 0.68, (b) 37.8°, (c) 1.75

P8.9   (a) 46.5°, (b) 0.78

Chapter 9

P9.9   (a) 54,545.4 mm^3/min, (b) 97,297.2 mm^3/min; ratio $= 1.78$

P9.11   273%

P9.13   (a) 37 min, (b) 16,215.4 mm^3/min

P9.15   (a) 30041.5 mm^3/min, (b) 1.27 min

Chapter 10

P10.7   2.42 min

P10.9   For each division on the job (work) the crank will move through 15 holes on the 18-holes index circle on the index plate by using the Plate # 1 of the *Brown and Sharpe Company*

Chapter 11

P11.9   1.6 h

P11.11   31 μm

Chapter 12

P12.1   925,423.7 K/s

P12.3   (a) 3.75, (b) 2.13 mm^2

P12.5   (a) 180, (b) 0.0055 in

P12.7   0.54 g/cm^3

Chapter 13

P13.7   (a) $\gamma_{SL} = \gamma_S - (0.98)\, \gamma_L$; the best wetting, (b) $\gamma_{SL} = \gamma_S - 0.5\gamma_L$, no wetting

P13.9   114 MPa

P13.11   4.1 kJ/mm

Chapter 14

P14.9   (a) 10^8 poises, (b) No, because 10^8 poises viscisity is too high to be formed; to be in the working temperature range, viscosity should be around 10^4 poises

P14.11   1.78

P14.13   Fused silica, because it has the highest annealing point

P14.15   97.5 cm

Chapter 15

P15.9   1.76 GPa

P15.11   0.45

P15.13   0.512 /min

P15.15   13.88 g/cm^3

Chapter 16

P16.7 0.366×10^{-5} mm

P16.9 0.135 mm^3/s

P16.11 8 min

P16.13 100 mm^3/min

Chapter 17

P17.9 $\begin{bmatrix} 5 & 6 \\ 6 & 7 \end{bmatrix}$

P17.11 4667 pulses

P17.13 105 Hz

P17.15 G02 X5.8 Y7.3 R12 F130

P17.17 N40 G01 X17 Y8.2 Z4 T05 S700 F130 M09

Chapter 18

P18.5 (a) 5.25, (b) 3.412

P18.7 61.8%

P18.9 0.212 mm

Chapter 19

P19.7 (a) 20,000 units, (b) $300,000

P19.9 $30.8 million

P19.11 $30,199

P19.13 0.95 hours

P19.15 (a) 126 units, (b) 16, (c) 16 years, (d) $632, (e) 40 units

Index